Continuous Improvement, Probability, and Statistics

Using Creative Hands-On Techniques

Continuous Improvement Series

Series Editors:
Elizabeth A. Cudney and Tina Kanti Agustiady

PUBLISHED TITLES

Affordability: Integrating Value, Customer, and Cost for Continuous Improvement
Paul Walter Odomirok, Sr.

Design for Six Sigma: A Practical Approach through Innovation
Elizabeth A. Cudney and Tina Kanti Agustiady

Continuous Improvement, Probability, and Statistics

Using Creative Hands-On Techniques

William Hooper

CRC Press
Taylor & Francis Group
Boca Raton London New York

CRC Press is an imprint of the
Taylor & Francis Group, an **informa** business

All graphs produced in Minitab Statistical Software. Please visit www.Minitab.com for details.
All photos shot at GAALT Studios, Granger, Indiana.

CRC Press
Taylor & Francis Group
6000 Broken Sound Parkway NW, Suite 300
Boca Raton, FL 33487-2742

First issued in paperback 2020

ISBN-13: 978-1-138-03507-2 (pbk)

Library of Congress Cataloging-in-Publication Data

Names: Hooper, William (Writer on statistics)
Title: Continuous improvement, probability, and statistics : using creative hands-on techniques / William Hooper.
Description: Boca Raton : CRC Press, 2017. | Series: Continuous improvement series | Includes bibliographical references.
Identifiers: LCCN 2016049741 | ISBN 9781138035072 (pbk. : alk. paper) | ISBN 9781315269498 (ebook)
Subjects: LCSH: Statistics--Study and teaching. | Probabilities--Study and teaching.
Classification: LCC QA276.18 .H66 2017 | DDC 519.5--dc23
LC record available at https://lccn.loc.gov/2016049741

Visit the Taylor & Francis Web site at
http://www.taylorandfrancis.com

and the CRC Press Web site at
http://www.crcpress.com

Dedication

This book is dedicated all students that have struggled to learn data analysis, statistics, or continuous improvement.

Contents

Part III: Introduction: Data, statistics, and continuous improvement via the experimental helicopter

Preface: Learning data, statistics, and continuous improvement another way

I remember it well. During a grad-level course in statistics, the professor whose name will remain anonymous, described the hypothesis testing method and told everyone to memorize the following for the exam: "Failure to reject the null hypothesis when the null hypothesis is false is a beta error and rejection of the null hypothesis when the null hypothesis is true is an alpha error." After class, several of us were exchanging notes when one student said to think of a courtroom trial as a better example as to what the hypothesis testing method is really saying. Intrigued, one person said, "Can you expand on that?" "For an alpha error, think of Nelson Mandela, and for a beta error, think of Al Capone." This opened up an entire discussion on how this almost 90-year-old theory can be used for experimentation and continuous improvement—not just memorized for the exam. How understanding that alpha error is falsely determining there was a change in a process when there was none, and beta error was the failure to recognize the change in the process when it really happened. How proper understanding of this method alone can begin to ignite the passion for continuous improvement in every machine operator, administrative clerk, plant manager, and leader.

After years of consulting by traditional methods, in 2011, I dedicated my life to a teaching method that would change the way the world sees data and probably more importantly taught data analysis and probability theory. What if instead of traditional lecture and test, probability could be learned by card and coin magic? What if the art of juggling could be used as a training technique in data analysis and statistics? What if the experimental helicopter could be used for teaching such concepts as the t-test, ANOVA, and design of experiments? What if 3D imaging could be used to visualize cube plots critical to understanding design of experiments?

This book is dedicated to all those who have struggled with the concept of statistics, have a genuine fear of data, and think the world of continuous improvement and experimentation is designed for a minor few and not for the masses. This book begins to answer the question; why can't every operator, technician, student, manager, and leader understand the fundamentals of data and the science of data analysis for incremental and many times breakthrough in continuous improvement?

I thank all my students from the continuous improvement courses over the past 10 years for their inspiration of this book and the hope that these methods will launch a new method of teaching and instructing in the science of continuous improvement.

William Hooper

Acknowledgments

Writing a book is about combining lonely effort with group therapy; I am forever grateful for those who supply the latter. Here is a partial list.

A special thanks to the editor, Kate Jakubas. We should all be so lucky to have her skill available for a project like this. Her attention to detail combined with questions regarding content on each chapter is what made this book. Could not have done it without her.

Also, special thanks to professional magician Ben Whiting. The concept behind many of the techniques in Chapters two and three can be traced to one-on-one instruction from Ben in 2010. While learning the card tricks, I started using them the next day in class to teach probability theory. That turned out to be a huge hit and a learning technique for the students, starting the process that eventually led to writing this book. Ben is a rare find in the magic world—a terrific entertainer, innovator, and a great one-on-one instructor.

Special thanks to professor Elizabeth Cudney of the University of Missouri S&T. Dr. Cudney attended my presentation before the 2014 American Society for Quality's World Conference in Dallas, Texas, when the concept of using card magic to explain probability theory was first introduced. That started the discussions with Dr. Cudney on the concept of writing a book based on that talk. She has been a terrific inspiration and coach through the book-writing process.

The use of juggling as a technique to teach continuous improvement was first brought up by a past student in 2015. My son Todd Hooper, who codesigned and presents the "Juggling for Creativity and Teamwork" workshop, and past Pro juggler Al Eisenhour, inspired the concept behind the use of juggling to teach continuous improvement. Juggling is an amazing, life-long learning process that few have started, let alone mastered. And as I hope all can see from Chapters five through eight, the sport of juggling follows very nicely the continuous improvement process.

The card trick in Chapter ten was inspired in part by seeing a video of the brilliant magician Wayne Dobson perform a similar trick, *Toss-out Deck*. Wayne Dobson performed this trick from a wheelchair while suffering from multiple sclerosis. He is now championing hands-free magic.

If there ever was an inspiration for using magic as a communications channel, it would be Wayne Dobson. To watch his amazing audience skills, follow him at www.dtrik.com.

Some of the card illusion techniques were an inspiration from working as a close-up magician for Open Heart Magic in the adolescent isolation sections at several Chicago area hospitals. Reducing the anxiety and putting a smile on a face for a child going through chemotherapy was a gift. Teaching Michael, who had a severe muscular disorder and who I could only talk to through visual language, was something to see. His reaction to the Dai Vernon's ambitious card routine was priceless, and the reaction from his parents, I will never forget.

Many thanks to juggling historian David Cain for the help with Chapter five on the history of juggling. The photos of Ronald Reagan, Barry Bakalor, and Alan Greenspan all came via David as well as the list of famous people who juggle.

Chapter four on Bayesian statistics was inspired by the failure of justice in the Sally Clark case. I hope everyone learns from the shortsightedness in this case and the damage that can occur from real-life alpha errors.

And finally, the true champion for this project—my wife of 38 years and best friend Maude. I could not have done it without her. She truly is the love of my life and inspiration for this lifelong endeavor.

Thanks to all.

Bill Hooper (spreading the love of data, statistics, and continuous improvement through card magic, juggling, and other innovative means).

Author

Bill Hooper is an independent consultant specializing in data-based productivity and quality improvement for small- and mid-sized companies. He holds an undergraduate degree in Engineering from the University of Michigan and an advanced degree from Indiana University, in addition to certifications from the American Society for Quality (ASQ) for Six Sigma Master Black Belt, Reliability Engineering, Quality Management, and Quality Engineering. Bill has implemented over 100 designed experiments for multiple industrial and service-based clients over the past 20 years, but likely is best known for teaching a series of innovative courses on data, statistics, and design of experiments throughout the United States, Canada, Africa, and the Middle East.

Bill is also a trained close-up magician and a performing juggler, best known for creating with his son, Todd Hooper, the workshop "Juggling for Creativity and Teamwork." The use of juggling as a training method for continuous improvement is from that workshop and also from teaching hundreds of children and adults to juggle. Bill previously volunteered for the Chicago area nonprofit corporation Open Heart Magic, an organization that specializes in the use of close-up magic to accelerate the healing process for hospitalized pediatric patients. Many of the techniques used in his courses were initiated while volunteering at Open Heart Magic.

Bill is a speaker and keynote speaker at various technical and nontechnical conferences on the use of card magic and juggling to teach data, statistics, and continuous improvement.

Learn more about Bill's unique courses and presentations at www.williamhooperconsulting.com.

part one

Simple steps to making probability interesting

chapter one

The science of learning

Make it interesting or shocking*

We all remember the first time we learned how to drive a car: It took maximum concentration to drive around the neighborhood as mom or dad sat in the passenger seat while the brain kept updating: "not too fast; adjust left; woops too far, adjust the other way; Ok give it gas; not that much; oh no, another car is coming at me; concentrate; the intersection is up ahead, slow down; I think I am getting it…" According to Robert Bjork, a psychologist at the University of California, Los Angeles, there are three concepts that are important for remembering events such as the first significant time at anything. These will be covered later in more detail, but for now think of them as useful, relevant, and interesting/shocking. Do we remember that incident? Most likely it is—highly useful, relevant, and interesting or sometimes called shocking. Bjork's findings state the greater the total of the three, the deeper the learning, such as having deep grooves in an album or more hard drive utilized. This leads to the statement of "it's just like driving a car" or "just like riding a bike." The findings are that we maximize the grooves in the album by the sum total of useful, relevant, and interesting/shocking.

Why we still remember where we were and what we were doing on 9/11

So why do we still remember where we were during 9/11, or, for those old enough to remember, when President John Kennedy was shot?

This does not mean we actively remember everything we learn. It depends again on whether the information is useful, relevant, and interesting/shocking. According to Robert Bjork, most of what we learn is there forever (this is hard to believe). My first phone number—still there. My first home address—still there. My first girlfriend's first and last name—still there. The first date with my wife—still there. The brain has the ability to hold a million gigabytes or three million TV shows. Unbelievable—but for most of us we really never run out of storage space. The problem is getting it into the grooves. Once it is there, under normal circumstances, it never leaves. Once it is in the grooves, according to

* A major portion of the chapter was obtained with permission from Carey (2014).

Bjork, there are several methods to enhance recall, but if it does not get in the hard drive, recall is difficult, if not impossible.

I'll offer an example from my own life. I used to work at an exclusive country club in high school. I was a huge fan of golfers at the time, as I played often. During one weekday, as I was working in the restaurant, I looked up at one of the tables and there was the legendary golfer Jack Nicklaus. I did nothing more than stare at him while I cleaned the tables—a childhood idol, sitting meters away. There was no meeting, no autograph, not even a glance back from him toward me. How could I still remember it years later?

I remember this incident because it was a combination of useful, relevant, and interesting/shocking. The greater the usefulness, relevancy, and interest, the greater the burn and deeper the grooves. But wait a minute, what about my seventh-grade science teacher who I had for an entire 8 months? Wasn't that class useful, relevant, and interesting/shocking? Why can't I remember a thing from that class? Here's why: Useful?—Ok, Relevant?—Ok, but interesting/shocking?—Not really. Probably the poorest class with a teacher who had little interest in being there. Again, the three concepts:

Useful—Ability to be used for a practical purpose or in several ways
Relevant—Closely connected or appropriate to the matter at hand
Interesting/shocking—Arousing curiosity; holding or catching the attention

Why do we all remember where we were on 9/11, or when John Kennedy was shot? Go back to the three combinations, and it may be explained why. All of those incidences likely scored high on useful, relevant, and interesting/shocking.

So, what was the difference between my Jack Nicklaus experience and an encounter with Bobby Nichols (another pro golfer from back then who was the pro at the club house and I saw daily for a year)? It was interesting/shocking. Go back to the original concept—it is the combination of useful, relevant, and interesting/shocking. The more interesting/shocking the encounter, the more it is burned into memory and the deeper the grooves. I challenge the reader to go back in time and recall an incident from years ago. Rank the three categories on a scale of 1–10 with 10 being the highest. Multiply the three numbers together and come up with an index number from 1 to 1000. This method is used in the concept of Failure Mode and Effects Analysis (FMEA), and it works here as well. Even with equal weights between the three categories, what happens to the three categories when multiplied together? Quite significant on the impact of learning.

But how does this relate to statistics, data, and continuous improvement? Here is my own personal story on this area. Recall the Pythagorean Theorem from high school math class—very useful in many settings. But let me explain how it was explained to me, and why I can remember it

40 years later. Again, remember Bjork's concept: the greater the combination of useful, relevant, and interesting/shocking, the greater the storage and likely the greater the retrieval.

In 1972 when I was taking the class in this subject, one of the smarter math instructors had us work through the derivation of the Pythagorean Theorem, following how the theorem was probably developed for the first time. Some of us may remember the formula $A^2 + B^2 = C^2$.

How relevant was this equation at the time? Let's put the three categories on a rank scale from 1 to 10, with 1 being totally useless (I would think that some of the calculus classes on integration turned out to be a 1) and 10 being used in everyday life (although some would argue differently, addition of two single digit numbers on a scale of 1–10 is probably a 10 as I may not be able to get change at the grocery store if I did not know that). So for the Pythagorean Theorem, maybe a 6 for relevancy (I needed it to pass the midterm but not much more), a 2 at best for useful, and maybe a 2 for interesting/shocking. The URI index (useful × relevant × interesting or shocking) number of 24 is not going to penetrate the grooves on the storage memory much at all—at least not past the midterm in 2 weeks. Say hello to Ben Weinstein, ninth-grade algebra teacher. Take a look at Figure 1.1 and work through the derivation of the formula.

a. The area of the entire square = sum of all interior parts.
b. C^2 = the area of the inside square + the sum total of the four triangles.

$$C^2 = (A \quad B)^2 + \frac{1}{2} A \times B \times 4$$

Work through the math and arrive at $C^2 = B^2 + A^2$.

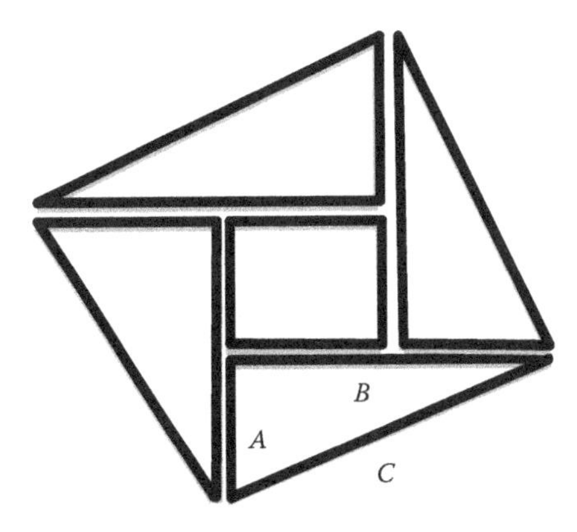

Figure 1.1 Diagram to assist in deriving the Pythagorean Theorem. (From Carey, Benedict, *How We Learn: The Surprising Truth about When, Where, and Why It Happens*. 2014, Random House, London. With permission.)

All those who enjoy challenges, come back to this picture the next day and work through the derivation. How hard is it to derive the formula? Now, come back 6 months later. Remember the picture and derive the formula. Now come back 1 year later and repeat the same process. It worked for that ninth grader working at the country club, and it works on students today.

So what changed? Why is it now grooved in memory when we would usually expect it to be lost 4 h after the midterm? Again, the combination of useful, relevant, and interesting/shocking. If the usefulness did not change (still a 6), and the relevance did not change (still a 2), what changed? For me, the concept of interesting/shocking increased from a 2 to probably a 9, and so the URI Index increasing from a 24 to 108—about a 4.5 times increase. Why the increase in interest? It changes memorization of a not-so-interesting formula to following a derivation process. I remembered the derivation process but did not remember the memorization (Table 1.1).

Try out the concept in other areas:

Learning to ride a bike for the first time (Useful—10, Relevant—10, Interesting/shocking—8, URI 800).

Table 1.1 Applying useful–relevant–interesting/shocking index to learning the Pythagorean Theorem

	Definition	Straight memorization of Pythagorean Theorem	Weinstein method: Deriving theorem
Useful	1: Great concept but has no useful application 10: Used in everyday life	6—Will use to get through homework this term	6—This did not change
Relevant	1: No relevance to anything I do 2: Highly relevant to my daily life	2—Not relevant to any part of my life	2—This did not change
Interesting/ shocking	1: Read in textbook or standard lecture 10: Fascinating how it was derived. Shocked when it was taught. Hands-on method when learned	2—Presented in standard lecture. Not interesting/ shocking	9—Worked through very unusual and unique derivation
Index	Useful × relevant × interesting or shocking (URI)	6 × 2 × 2 = 24. Not very sticky	6 × 2 × 9 = 108. Much stickier than before

Learning calculus for the first time (Useful—3, Relevant—2, Interesting/shocking—2, URI 12). Okay, that may change depending on how you use it. But I would think most high school students are in that ballpark.

Learning how to compute a baseball player's batting average (Useful—8, Relevant—8, Interesting/shocking—10, URI 640). Okay, this might not be for a 10-year-old in the Congo, where they probably have limited knowledge of baseball, but it is high for a 10-year-old in Japan or the United States.

Learning how to compute average and standard deviation for a machine operator (Useful—6, Relevant—6, Interesting/shocking—2, URI 72). Wonder why it is so hard to get industrial personnel interested in statistical process control (SPC)!

Learning how to find an interaction effect in a process through a Designed Experiment (Useful—8, Relevant—8, Interesting/shocking—2, URI 128). Okay, that is until you understand how to visualize the interaction term, and at that point the interest likely increases from a 2 to a 7. [That will be the topic of a future book dedicated to how to teach Design of Experiments (DOEs) for everyone from ninth grade to college.]

Learning that the infant mortality range of the bathtub curve for product failure has a decreasing breakdown frequency (Useful—8, Relevant—8, Interesting/shocking—3, URI 192). Until of course, the process of learning arouses the curiosity of the student by creating interest through alternative means.

Overview of the chapters

The student cannot apply data analysis and statistics to industrial or transactional processes and everyday life without having the confidence to do so. Critical to building confidence is increasing recall of the concepts themselves. The goal of this book is to introduce a process to learning basic statistical, data, and continuous improvement concepts that make the learning useful, relevant, and interesting/shocking. In the author's experience, the weakest of these three using traditional teaching methods is interesting/shocking. This book focuses on strengthening the interesting/shocking value of learning data analysis and statistics.

The book is set in three parts.

Part one: Chapters two through four. Card tricks and probability

Part one explains how to apply the concept of card magic to learning probability theory. Developed with the assistance of a professional magician,

Ben Whiting, several card tricks are used as demonstration tools. This will only be the start of the opportunity for an interested instructor, as there are thousands of card and coin tricks dating back to 2000 BC.

Chapter two covers multiplication theory of probability using two card tricks to communicate and demonstrate the basic concept of statistical dependence and independence. *Chapter three* discusses a subcategory of probability and uses a third card trick to demonstrate an event with probability of occurring less than winning almost any state lottery system. This one will demonstrate the math behind the theory. A fourth card trick called the "phone number" will be used to demonstrate the probability of randomly identifying a person's phone number. *Chapter four* describes Bayesian statistics. This relatively confusing and little-used technique in most industrial or process applications will be demonstrated by a coin trick and by other simplified examples. Hopefully, this chapter, helping with increased understanding of Bayesian statistics, will inspire increased use of this process.

At the end of *Part one,* some of the concepts of teaching probability theory by card magic will be available for teachers of mathematics, statistics, probability, Six Sigma, or continuous improvement.

Part two: Chapters five through ten. Juggling and statistics

Part two covers basic statistical and continuous improvement concepts illustrated by the use of juggling. The concept of combining statistics and juggling is new to the math and science world, but some of the most creative educators and scholars are jugglers, including the inventor of the computer and a very famous US President.

Chapter five explains the basics of juggling—how to juggle—from my own workshop on the use of creativity and teamwork for juggling. This chapter will have limited math and science but is high on learning of a new skill or hobby. *Chapter six* covers the development of the normal distribution and standard deviation for the number of tosses to failure. This concept is the start of the mathematical modeling of the process of juggling. *Chapter seven* discusses the basics of SPC as taught by juggling. This interesting/shocking method of measuring the process development of juggling will increase understanding of how to properly utilize SPC for continuous improvement. SPC will be developed for the number of tosses to drop as a simulation of process improvement. *Chapter eight* describes juggling and equipment reliability, specifically the use of the bathtub curve to explain the break-in period and the wear-out period using number of tosses to failure. *Chapter nine* explains a very basic DOE for the juggling process followed up with a very simple regression model based on the DOE. *Chapter ten* focuses on a card illusion using a 5-factor, 32-run, full-factorial DOE as the model. This card illusion will demonstrate the concept of full factorial DOEs for a large number of factors.

Part three: Chapters eleven through thirteen. Experimental helicopter for continuous improvement

Part three will be on the use of the experimental helicopter for continuous improvement. The use of an experimental helicopter is not new, but its use to explain basic statistical concepts and advanced DOE with adjustable racks to modify the process during the experimental runs is new.

Chapter eleven will be on the use of the experimental helicopter to explain hypothesis testing and the ramification of excess variation on the traditionally difficult-to-understand Beta and Alpha errors. *Chapter twelve* utilizes the experimental helicopter to demonstrate one of the most powerful and underutilized Designed Experiments ever developed—the 5-factor, 2-level, 16-run half-factorial DOE. *Chapter thirteen* focuses on the optimization process for an experimental helicopter after running the DOE in Chapter twelve.

Part four: Chapters fourteen and fifteen. Making data and statistics fun and interesting

Chapter fourteen puts it all together and starts the process of determining how best to utilize this method in a new education method—one that now has a concept of interesting/shocking added in, leading to a major increase in the URI Index.

The aim of this book is to not only cover all areas of statistics, data, and probability theory but to assist a new generation of instructors in utilizing a set of effective techniques for data analysis, statistics, probability, and more recent areas such as Six Sigma or Lean Manufacturing.

It is my hope that using instructional methods such as juggling and card magic will lead to an increase in understanding of data and statistics, inspiring a generation of continuous improvement experts who otherwise would never have been found. Being inspired, this next generation will lead breakthroughs in advanced manufacturing, medical research, service industry, information technology, or other areas.

Bibliography

Bjork, Robert A.; Bjork, Elizabeth Ligon (eds), *Memory*. 1996, Academic Press, San Diego, CA.

Carey, Benedict, *How We Learn: The Surprising Truth about When, Where, and Why It Happens*. 2014, Random House, London.

Pashler, Harold; Carrier, Mark, Structures, processes and the flow of information. In Bjork, Robert A.; Bjork, Elizabeth Ligon (eds) *Memory*, 4–25. 1996, Academic Press, San Diego, CA.

chapter two

The use of two very basic card tricks to explain probability theory

Objectives

- At the completion of this section, instructors should be able to utilize two very basic card illusions to explain the multiplication rule of probability theory.

The basics of the multiplication principle for probability theory

Let's start with the basics:

$$P(A \times B) = P(A) \times P(B|A)$$

Or standard written form: Given two events, what is the probability of both occurring in succession? Notice the verbiage: The first event precedes the second, or there is lack of independence between events. If both events are independent, the formula changes to the following:

$$P(A \times B) = P(A) \times P(B)$$

The second term changes to an independent event, such that it does not depend on the first occurrence (i.e., independent events).

The concept is used throughout equipment reliability, predicting process efficiency, determining quality products output, etc. But, how hard is it to understand the difference between the two concepts? Let's first utilize two simple card tricks to explain the concept. This will be followed by practical applications of the probability theory.

Classroom activity 1: Teaching the multiplication principle by card trick—The aces and kings card trick

Before starting, procure a deck of normal playing cards and practice some very basic concepts with the students:

How many cards are in the deck? (52)
How many kings are in the deck? (4)
How many aces are in the deck? (4)

What is the probability of shuffling the deck and randomly pulling a king? This should be relatively easy as there are exactly 4 kings and exactly 52 cards. It is 4/52 or roughly 7.7% or let's call it 8%. Seems like a simple concept built on the concept of shuffling the deck and randomization. But what if this method could demonstrate the probability of two kings, three kings, or the extremely rare four kings followed by four aces?

The step-by-step mechanics of how a simple card trick can greatly enhance the learning process

Step 1: Preload the deck

Take the deck of 52 cards and prestack the deck with the four aces on top and the 4 kings at the bottom, prior to the in-class demonstration (Figures 2.1 and 2.2). For best effect, return the deck to the pack, although this step is optional.

Step 2: Bring up to the front a member of the class or audience

Bring up a member of the class or audience to perform the trick with you. This person can be randomly chosen, but make sure the candidate is able to shuffle a deck and has some familiarity with card playing and basic poker hands.

Step 3: Optional fake shuffle #1

Remove the deck from the pack and perform a fake shuffle. What is meant by a fake shuffle? Shuffle only the middle section of the deck, leaving the top four preloaded aces and bottom four kings in the same spot. This can be done as shown below in Figures 2.3 and 2.4. Split the deck in half and start by riffle shuffling the deck except the top 4–10 cards, leaving the top 4 aces in place. Drop the bottom 4–10 without shuffling, leaving the 4 kings in place. This will give the impression that the deck was shuffled, which it was except the first and last 4–10 cards—the 4 aces and 4 kings

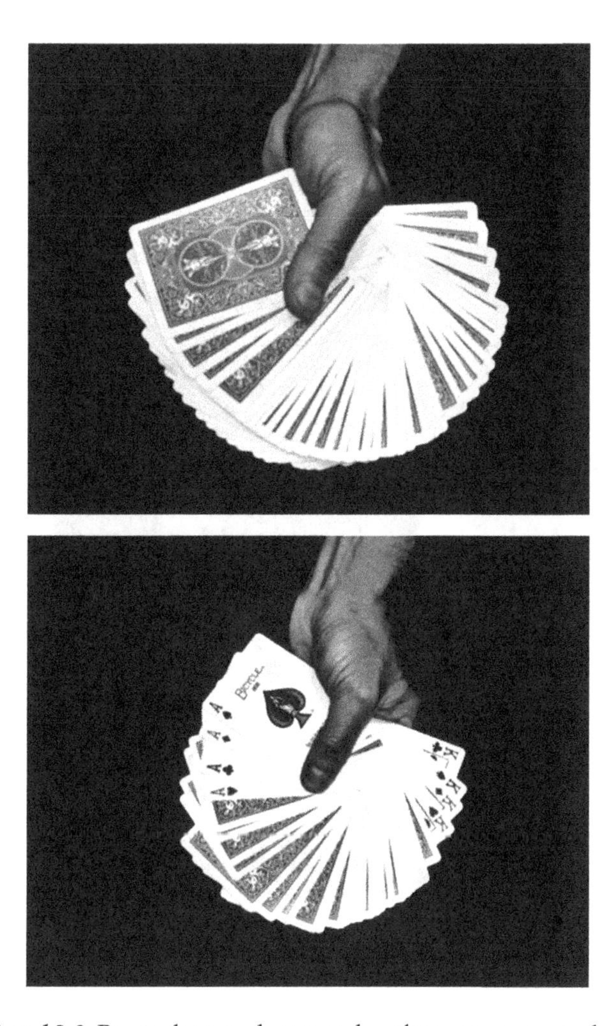

Figures 2.1 and 2.2 Prestack turned over and as they appear to students.

positions are not moved. This technique is almost impossible to detect by the students or the audience. Nonetheless, the trick can be done very effectively without this step.

Step 4: Strategically have the student cut the deck

After the optional fake shuffle, have the student cut the deck nearly in the middle of the deck. This step is critical to the execution of the trick. Best to state, "Because I am going to go through how to shuffle cards, it is best to cut the cards near the middle of the deck." They must not cut the deck at the top four aces or the bottom four kings. This is rarely done, but it will

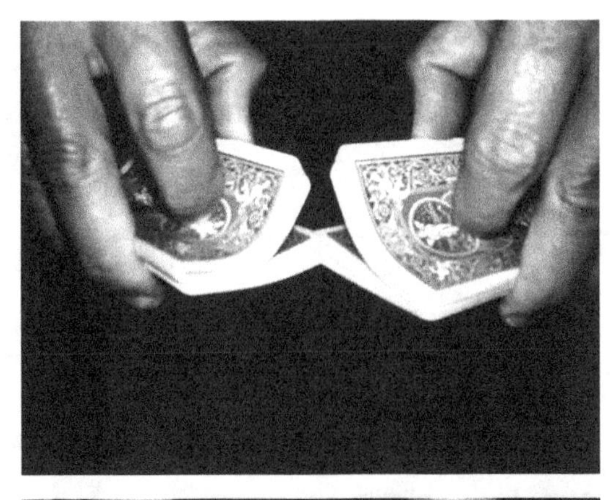

Figures 2.3 and 2.4 Standard shuffle and shuffle holding kings at the bottom and aces on top.

ruin the effect. At the end of this cut, the four aces will be on top of the student's partial deck and the instructor will have the kings at the bottom of the deck, as shown in Figure 2.5.

Steps 5A–5D: The fake shuffle. Moving the four aces and the four kings to the bottom four on the table in what is believed are fully shuffled cards

Shuffling cards has a long history. The standard riffle shuffle was started only around the start of the century. During this next step, the instructor moves the four aces and four kings strategically to the bottom four cards on the table under the disguise of elaborating on the history of shuffling cards. Here is how it is done.

Figure 2.5 Cards cut with instructor's kings at the bottom and student's aces on top.

Figure 2.6 Moving kings from bottom of instructor's half to new stack.

Step 5A: Move your four kings from the bottom of the deck to the bottom of a new stack on the table

Explain the history of different methods used over the years of shuffling cards. The dialogue typically used is as follows: "We are familiar with the standard shuffling methods such as the riffle method or overhand method. I will get to those methods in a moment." Early shuffling techniques included such methods as just reversing the deck [this is when you and only you pull the bottom four cards from your stack to the table, creating a new stack (see Figure 2.6)]. This step is critical as the student cannot move her/his cards to mirror you, because their aces are at the top of their pile. The purpose behind this step is to move the four kings from your bottom deck to the new stack on the table while the student's deck remains unchanged.

Step 5C: Move the student's stack with the four aces on top to her/his new stack

Your four kings have been moved to the bottom of the new stack. This position is critical for the next step. But the student has not moved or created their stack. The next dialogue is critical to the student shifting their aces to their newly created bottom stack. The typical dialogue: "Another method of card shuffling utilized years ago was the reverse order technique. Very simply, reverse order the top cards—why don't you do this with me." At this time, you both move the top four cards from the top to the table—you with the cards on top of the four kings, while the student has started a new pile on the table with the four aces at the bottom (see Figure 2.7).

Step 5D: Shuffle all cards except the cards that have been placed on the table

This step is the easiest to follow but probably the most critical. We are going to waste some time and perform some steps that have no bearing on the trick. There is a concept used in product investigations that is a close analogy to this step: the longer the time that elapses after an event and the more events that happen between the investigation and the event, the less likely the event details are to be recalled. This next stage is designed to make the student forget that the entire deck was not shuffled (the four aces are at the bottom of the pile, while the four kings are at the bottom of the instructor's pile). Both the student and the instructor will shuffle the cards remaining in their hands, leaving the cards on the table untouched and hopefully forgotten. Have the student take her/his remaining cards

Figure 2.7 Moving the top four cards to the student's new stack and the instructor's existing stack of a minimum of four cards on the table.

and perform conventional riffle and/or overhand shuffle on the cards. The student and instructor begin this activity starting independently. After a shuffle or two, give the student your instructor stack of cards and have him/her shuffle the two together. While the student is shuffling the deck, talk about the history of randomization, other card shuffling techniques, or any appropriate topic. The ideas once again are to make sure the student forgets that the entire deck was not shuffled (i.e., the fake shuffle) as they will be shuffling all cards except those on the table, which have the four kings and four aces at the bottom (see Figures 2.8 and 2.9).

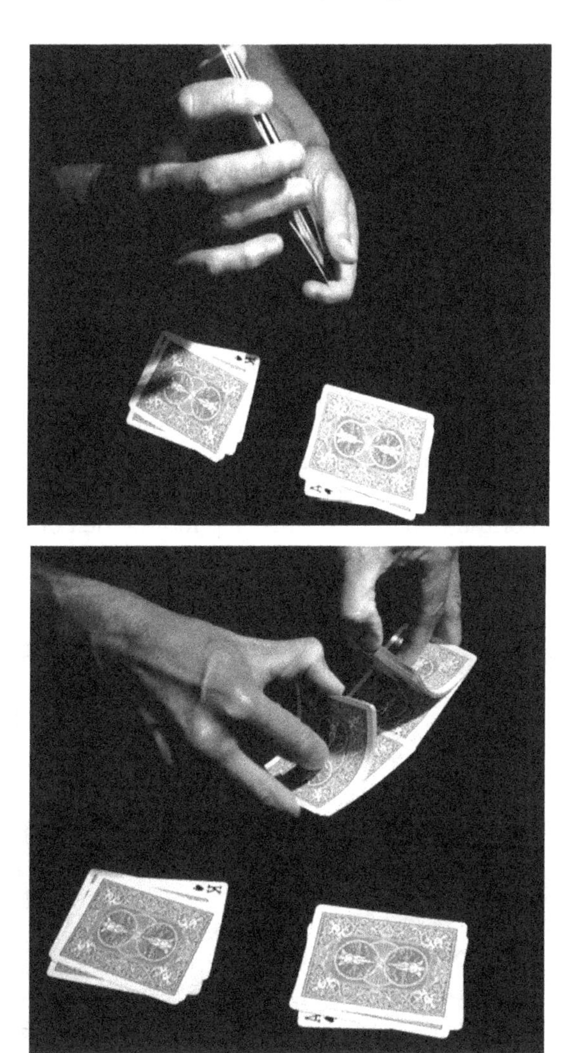

Figures 2.8 and 2.9 Shuffling the entire deck with the four cards on the table.

Step 6: The fake shuffle continued

Once the cards are shuffled multiple times, have the student give you approximately half the cards back and both of you place the shuffled cards on top of the cards on the table. This step is one more full randomization exercise to seal the deal that the deck was fully shuffled when actually they never realized that the first four cards were never included in the shuffle. This step essentially creates an approximate half of deck for you with the kings at the bottom and the student with their half and the aces at the bottom (see Figure 2.10).

Step 7: Form four piles of cards on each side of the table from the two halves

In this step, create four piles from the single instructor pile, each with a king on top, and four piles from the student pile, each with an ace on top. Take the deck that they think was shuffled and begin dealing the cards as if this was a four-person poker game. Have the student join you, following your motions to deal out their own pile of cards into four piles. The four aces and four kings are now still at the bottom of the half decks. Have the student or someone in the audience shout out the number—one, two, or three. Place one, two, or three cards on each of the four stacks depending on the number shouted out. Do this for about one round making sure the bottom four cards (the aces and kings) are not touched. At that point, stop the counting for time reasons and proceed to deal out the remaining cards, one at a time on top of the four stacks. This step moves the bottom four cards to the positions of the top card on the four stacks—the four kings for the instructor and the four aces for the student (refer to Figures 2.11 and 2.12).

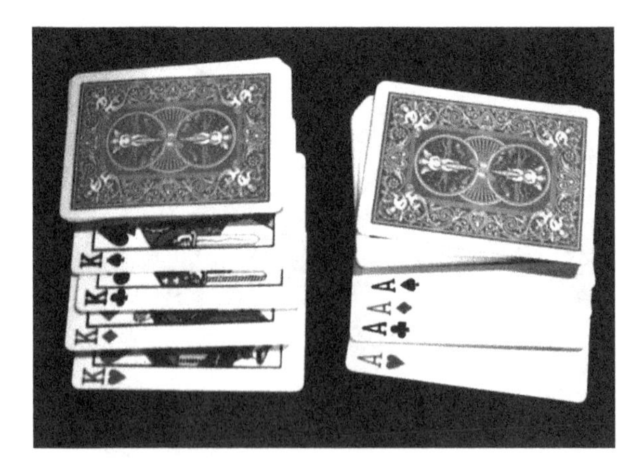

Figure 2.10 Two piles complete on the table.

Figures 2.11 and 2.12 Dealing the four stacks on the table.

At this point, the cards are set with the four kings on the top of your four stacks and the four aces on top of the student's four stacks turned over facing down. The stage is set for the use of the probability theory and shocking finale.

Card trick finale: Using multiplication principle of probability theory

Start with the four kings. The dialogue for this section can be customized for each specific circumstance. Ask, "What is the probability of the instructor randomly turning over a king?" That was computed earlier—it is roughly 8%. The dialogue can be customized to the technical ability of the students or audience. After having the class agree on the probability of turning over a king, reveal one of the top cards, which will be one of the four kings. Next ask the audience, "But what about a pair of kings after turning over the first king?" This follows the multiplication principle as this event has a conditional probability after turning over the first king.

$$P(A \text{ and } B) = P(A) \times P(B \mid A)$$

$$P(A \text{ and } B) = P(\text{first king}) \times P(\text{second king given that the first was exposed})$$

$$P(A \text{ and } B) = \frac{4}{52} \times \frac{3}{51}$$

$$P(A \text{ and } B) = .0045 \text{ or } .54\%$$

Not very large but possibly could happen. This is an excellent time to review other statistical probabilities. This probably is actually not outside a standard control chart upper or lower control limit. Great exercise commonly missed even by many experts in the field.

Typical examples of events that fall into this category might be as follows:

- Close but not there yet of a point outside the control limit on a statistical process control (SPC) chart.
- The probability of hitting a specific color and number on a roulette table—wonder why so many towns now have gambling casinos?
- The probability of a typical hole in one on a putt-putt course—this is why they entice you with free follow-up rounds of golf if you get it in. Great payback for not much advertisement.
- The probability of being injured in a car accident this year.

Next, return to your card piles and turn over another king—it does not matter which one as they are all kings. This is where it starts to get shocking. Turning over three kings is possible, but is it probable? Three kings? A rare event by the following multiplication principle. All these follow conditional probability as we are asking the probability of each event after new information becomes available.

$$P(A \text{ and } B \text{ and } C) = P(A) \times P(B|A) \times P(C|A \text{ and } B)$$

$$P(A \text{ and } B \text{ and } C) = \frac{4}{52} \times \frac{3}{51} \times \frac{2}{50} = .0002 \text{ or } .02\% \text{ or 2 out 10,000!}$$

Finally, reveal the fourth and last king. What is the probability of four kings being pulled from a shuffled deck? Same logical path.

$$P(A \times B \times C \times D) = P(A) \times P(B \text{ given } A) \times P(C \mid (A \text{ and } B)$$
$$\times P(D \mid (A \text{ and } B \text{ and } C)$$

$$P(\text{King} \times \text{King} \times \text{King} \times \text{King}) = \frac{4}{52} \times \frac{3}{51} \times \frac{2}{50} \times \frac{1}{49} = .0000037 \text{ or } 3.7 \text{ per every million opportunities}$$

However, it is close to the probability of a defect occurring at a Six Sigma level with a one and a half standard deviation process shift.

The odds of getting beat

So far, this example has been on the kings or the instructor's side of the four stacks. What about the other side? Remember now, the other side has the four aces on top. The student assistant and the audience do not know this. To solidify the deal for the multiplication principle, ask your student, "What are the odds that you will beat me? 10,000 to one, 1,000,000 to one?"

The multiplication principle is applied here for the audience to calculate these odds. There are now only 48 cards unturned, but no aces have been exposed. So under the assumption that the deck has been shuffled, each card will be turned over independently, and the only hand that will beat the instructor is four aces. They just happen to be the top four cards of the student's four piles. The odds of this happening are:

$$\frac{4}{48} \times \frac{3}{47} \times \frac{2}{46} \times \frac{1}{45} = .0000051 \text{ or } 5.1 \text{ out of a million}$$

It is interesting to note that the odds of hitting four aces increase if your opponent also has four of a kind. That rule applies not just to four of a kind, but also to many other probabilities in card playing. Your hand is not an independent event but is determined by the other hands showing, and the odds of a pair, three of a kind, etc., can increase significantly if you are in a high stakes poker game and another pair shows up on the board. Odds of getting three of a kind increase even more when three of a kind are on the board. Those who understand the game of Texas Hold-em understand that your odds of three of a kind go up significantly, when other pairs show up on the flop.

At this point, have the student turn over their top four cards. If all steps have been followed, the top four cards are the aces. In a standard game of five card straight—no draw, they just beat your hand of the four kings.

This will be the ultimate shock value and will significantly increase the "interesting/shocking" value in Robert Bjork's method of adding to the memory storage.

This concept will work for lack of independence, but what about independent events?

A card trick for teaching probabilities of independent events

What happens to the multiplication principle when the probability of the second event has nothing to do with the probability of the first event? This is called independence and changes the formula.

A review of the formula for dependence:

$$P(\text{A and B}) = P(A) \times P(B \mid A)$$

or the probability of A times the probability of B given that event A already happened.

But if the probability of B has nothing to do with A occurring, the formula changes to:

$$P(\text{A and B}) = P(A) \times P(B)$$

How to show the difference by card illusion

This is a classic trick first developed by Magician Dai Vernon and can be used for developing the concept of independence.

Step 1: Student selects and memorizes a card

No preload is required. This one is best done after the aces and kings trick (described earlier). Shuffle the deck and have the student remove a random card and memorize it. Let's call it the 6 of hearts for the remainder of this exercise (Figure 2.13).

Step 2: Deck split

Pull down the deck from the front, telling the student to stop at any point. When the student says stop, split the deck between your two hands at a stopping point (see Figure 2.14).

Figure 2.13 Student removing random card from the deck.

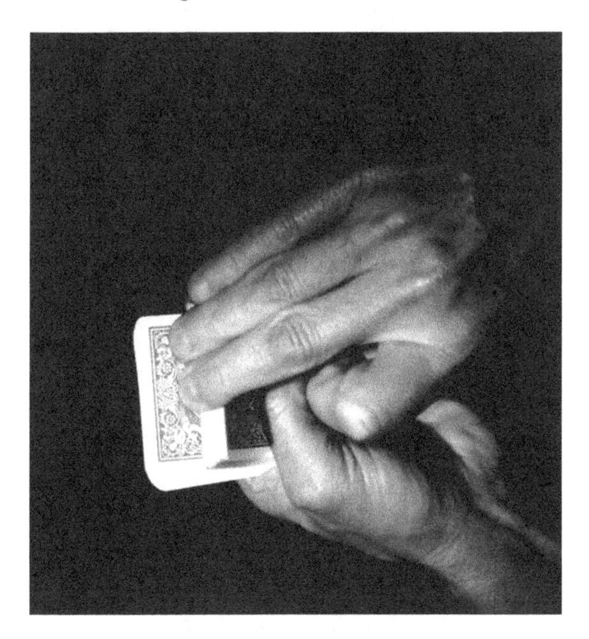

Figure 2.14 The deck being split at approximate half way point.

Figures 2.15–2.17 Returning the card to the top of the deck in left hand and creating what is called a pinky break.

Step 3: Card insertion and pinky break

Have the student insert the card on top of the deck in your left hand. Hold a break with the left hand little finger above the position of the card inserted and replace the stack from the right hand on top of the break (see Figures 2.15 through 2.17).

Step 4: Moving the target card to top of the pile

Pull off cards multiple times (3–5 groups of 5–7 cards) onto a separate pile (Figures 2.18 and 2.19) until the finger break is reached, and drop the rest

Figures 2.18 and 2.19 Off shuffling the cards down to the pinky break held card.

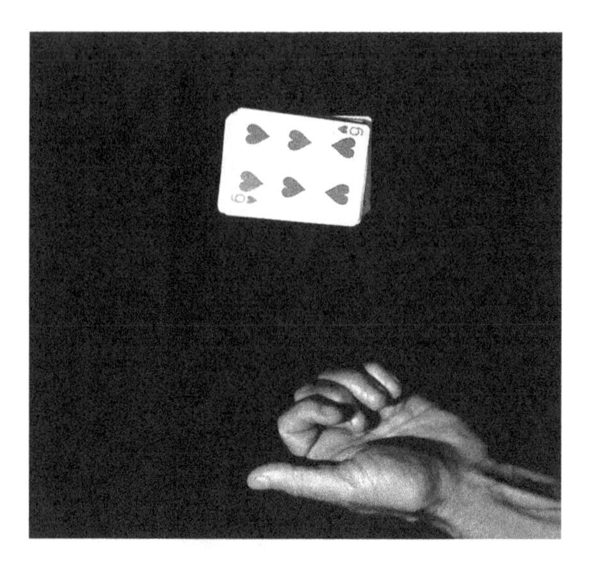

Figure 2.20 Dropping the rest of the deck below the pinky break from the bottom half to the top of the deck on the table.

of the pile on the new pile (Figure 2.20). This has the effect of moving the card just below the break—which was the card chosen by the student—to the top of the deck.

Step 5: False showing of card to student

Perform a double lift on the top two cards. This is best accomplished as shown in the following pictures, and has the effect of showing the second card down. The student will think you are showing the top card. For this example, the top card is the 6 of hearts (see Figures 2.21 and 2.22 for how to perform a double lift). The card chosen and shown will be a random card from the deck—the second card down from the top.

Step 6: Reveal

Have the student take the top card, which is the 6 of hearts. When called, have them turn over their card just handed to them (Figure 2.23). This will add to the shock value of the process, as they should be convinced the card in their hands is not the 6 of hearts.

Step 7: Returning the target card to the deck

Have the student return the 6 of hearts to the deck while you hold a pinky break at the location they return it to (Figure 2.24).

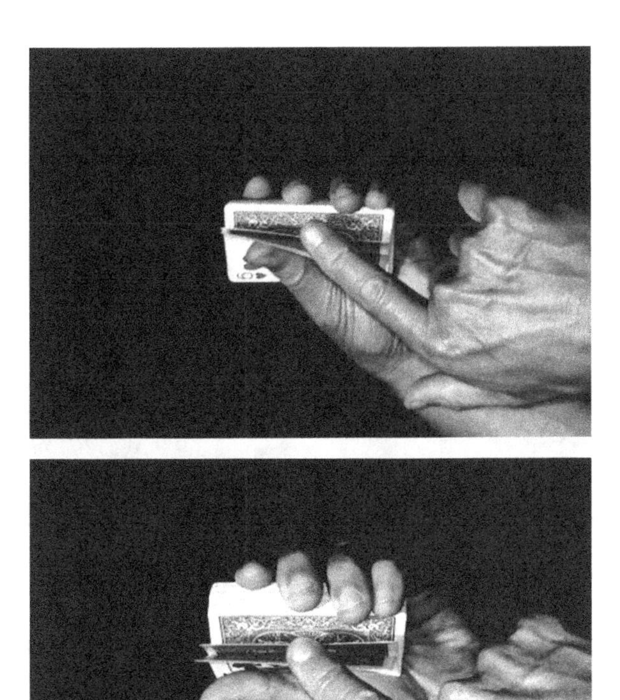

Figures 2.21 and 2.22 The double lift with the 6 of hearts exposed.

Figure 2.23 The student holding what he/she believe is a random card but is the 6 of hearts.

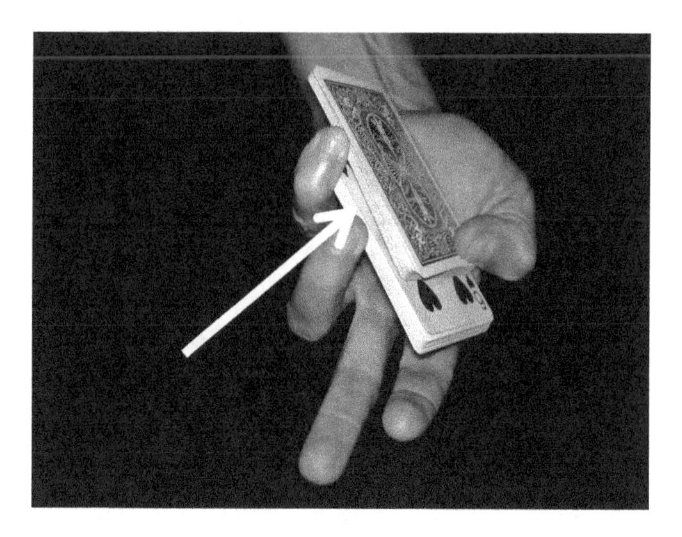

Figure 2.24 The deck cut at the 6 of hearts with a pinky break.

Repeat Step 7 four times as a method of explaining the concept of independence. Each time, return the card to the deck in a random position. Each time, pull the 6 of hearts. This is best done with four different students, thus preventing a specific student from possibly seeing the pinky break or double lift.

Comparing dependent and independent events

The probability of revealing the first king is always 4/52 or 8%. What if we had shuffled in between draws, like we did in Activity 3? The probability of drawing a second king is no longer .0045. It changes after the first card since the first draw was replaced and the deck reshuffled—therefore independent.

The probability of two kings when shuffling the deck in between draws (Table 2.1):

$$\text{P(King and King)} = \text{P(King)} \times \text{P(King)} = \frac{4}{52} \times \frac{4}{52} = .006 \text{ or } .6\%$$

$$\text{P(King and King and King)} = \text{P(King)} \times \text{P(King)} \times \text{P(King)} = .00046 \text{ or } .046\%$$

$$\text{P(King and King and King and King)} = \text{P(King)} \times \text{P(King)} \times \text{P(King)} \times \text{P(King)}$$

$$= \frac{4}{52} \times \frac{4}{52} \times \frac{4}{52} \times \frac{4}{52} = .000035 \text{ or } .0035\%$$

Table 2.1 Comparison between the two methods

	Not independent (card not replaced, the deck remains the same) (%)	Independent (replaced the card and reshuffled) (%)
One king	8	8
A pair of kings	.45	.6
Three kings	.02	.046
Four kings	.00037	.0035

Reality check: How it works in operations

Two machines are running in series with uptime of 90%. What is the probability of making shipments today?

This follows the multiplication principle with total independence with the formula as follows:

$$P(\text{A and B}) = P(\text{A}) \times P(\text{B}) = .9 \times .9 \times .81 \text{ or } 81\% \text{ probability of running}$$

Two machines are running in series and they share power. The power is not clean and therefore is determined by how many units are running. The more the running units, the greater the probability of a shutdown. Machine A has an uptime of .9, but machine B has an uptime of .8 if machine A is running. What is the probability of making shipments today?

This is the multiplication principle with dependent events.

$$P(\text{A and B}) = P(\text{A}) \times P(\text{B} \mid \text{A}) = .9 \times .8 = 72\%$$

Summary and potential next steps for instructors and students

The multiplication principle is critical to understanding probability theory. But, it is confusing and difficult for many first-time students to learn. Memorization will not help. Understanding of the process will aid in the students' development.

Author's note

In my experience, most attempts at teaching first-time students are an admonition to memorize the formula but not apply in their own workplace. Remember this concept from the first chapter: The more "useful, relevant, and interesting/shocking" the concept, the greater it is burned into the brain storage. The greater the storage, the easier the retrieval. Critical in the applications for continuous improvement in the workplace either in manufacturing or in transaction, the stronger the storage, the greater the likelihood of use in a real-life application. The card manipulation method not only aids in the learning of the concept, but also helps most students relate to the application in their own workplace. This has application for an entire manufacturing plant to a bank teller operation (compute the probability of a customer making it through the banking process when the transaction process involves waiting in line, bank teller computer wait time, printing receipt, etc.) to a single machine line operator. The computation can be used to compute uptime, project failure rates, rework operations, and for on-time shipments.

Chapter three will deal with a similar concept in probability and a card trick called the Phone Number developed by my colleague, Magician Ben Whiting. Before proceeding on, the foregoing concept is best covered in detail and trialed with family members or friends before a performance. By the third successful trial, the concept will be grooved in storage. As reviewed in Chapter one, the retrieval is strengthened by time delay between attempts and increased number of failures. Whichever way, the concept of recall and application will be strengthened for years by this method.

Bonus section: The process map

This section is done after the foregoing card trick. It is designed to strengthen the team-building process and investigation skills critical for product investigations.

What happened? This is a rare event. The odds of it happening are greater than 1 in 10 million (see Chapter three). What should we do next? Why did the rare event occur?

Students need to start the investigation process immediately, as time is the enemy of any process investigation. As a first step, create a flow chart of the process from beginning to end (Figure 2.25).

A process developed from the author's courses on how to perform a deep dive investigation after the flowchart begins with the question: What are the likely suspects at each step in the process? This is the first of two steps in the failure mode and affects analysis development. How likely did the step cause the rare event, and if it did, what are the odds that it caused the outcome?

Step 2—develop table and assign likelihoods to the events with the class (Table 2.2).

Recreate the highest probability of events. This should simulate the investigation process resulting in the recreation of the event. At the

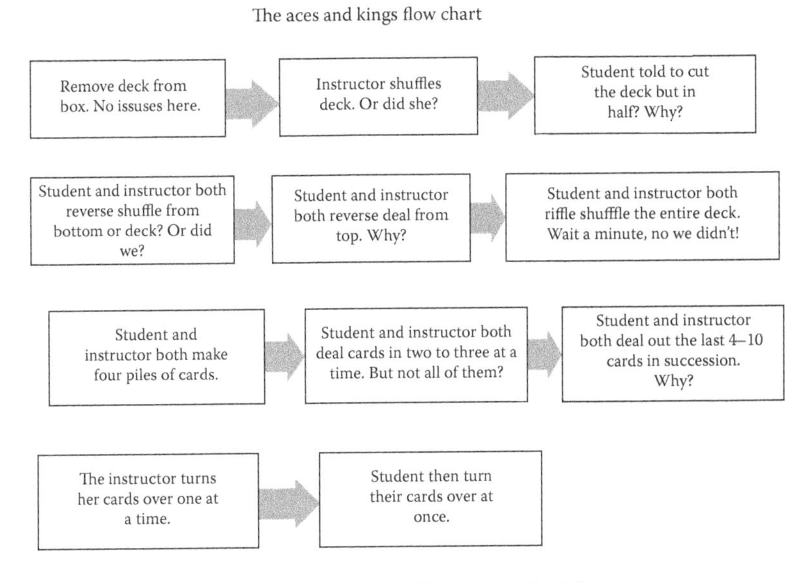

Figure 2.25 The flowchart of the aces and kings card trick.

Table 2.2 Typical ranking of each step in the process

Event	Likely to happen	Likely that it caused the kings and aces on top	Probability that it was culprit
Never shuffled the cards at the start	.1	.1	.01
The instructor somehow switched the cards around	.001	.1	.0001
The instructor did not shuffle all the cards—just some	.9 (we know now after the flow chart)	.9	.81
The student did not shuffle all cards—just some	.9 (we know now after the flow chart)	.9	.81
The cards were at the bottom of the instructor's deck all along	.5	.9	.45
It was a fake deck. There were only aces and kings in the deck	.1	.1 (even if it was, how did she keep the cards separated)	.001

completion of this step, this should simulate a typical five why's investigation process and provide an interesting and informative process to teach most investigation methodologies.

Summary and next steps

Used properly, this concept will solidify the area of multiplication theory of probability. But more important, by shock value of the card trick, probability will be grooved into the memory much deeper. This will lead to an increase in long-term recall.

Bibliography

Chen, Bill; Ankenman, Jerrod, *The Mathematics of Poker,* 14–16. 2006, ConJelCo LLC, Pittsburgh, PA.

Hays, William L., *Statistics,* 5th Edition, 130–135. 1986, Holt Rinehart and Winston, Inc., New York.

Menenhall, William; Reinmuth, James E.; Beaver, Robert; Duhan, Dale, *Statistics for Management and Economics,* 5th Edition. 1986, PWS Press, Boston, MA.

chapter three

Combinations and permutations by card magic

Objectives

- Review the concept of combinations in probability theory.
- Utilize the "phone number" card trick to demonstrate the concept of combinations.
- Understand how to apply combinations in the mathematics of poker and other games.

Overview

This chapter will expand on a specific area of probability—combinations. Many times utilized in introductory statistics courses, this section will utilize two additional card tricks to explain combinations. Once properly demonstrated, this concept will likely be remembered for a lifetime.

The concept of combinations and the phone number card trick

The following is a very simple and traditional computation example. There are three offices to be elected at large. Five people are on the ballot for the positions. In how many different ways can three positions be filled from the five candidates? Notice order does not matter, meaning there is no "first" or "second" position.

The solution follows the combinations formula.

Number of combinations irrespective of order is

$$\frac{N!}{(r!)(N-r)!}$$

So in this very basic example, the number of combinations, or ways to fill three positions from five candidates, is

$$\frac{5!}{3!(5-3)!} = \frac{(5\times4\times3\times2\times1)}{(3\times2\times1)(2\times1)} = 10$$

This example, or others very close to it, is the base example used in most introductory textbooks. Is it useful—does it have the ability to be used for a practical purpose? Probably. Is it relevant—is it closely connected to the matter at hand? Depends on the application, but likely. Is it interesting/shocking—does it arouse curiosity? Not really.

Go back to Chapter one, and recall the concepts of memory storage: Useful, relevant, and interesting/shocking. The higher the index value of the three, the higher the memory storage and the better the long-term retrieval. Useful and relevant. Maybe a 6 for useful and 4 for relevant, but for interesting/shocking? Maybe a 1 at best, unless the students are all running for office. Let's use a card trick, introduced for teaching by magician Ben Whiting, to demonstrate the concept and make this area more interesting and more memorable, thus improving the retrieval process. This trick will be called the "phone number" card trick. The concept of the phone number will be explained in the second half. The first exercise will be for poker play—specifically the odds of a flush hand in five-card poker.

The mathematics behind a flush in poker

What is the probability of a flush in a five-card no-draw poker game? For those unfamiliar, in a standard deck of 52 cards, a flush is being dealt 5 cards all the same suit. Example: 5-2-3-7-king of spades. Notice that order does not matter, and unlike Chapter two, since we are asking before the cards are turned over one at a time, the concept of conditional probability does not apply.

In a legal and fully shuffled deck of 52 cards, how many unique combinations of 5 cards are there? This follows the combinations formula:

$$\text{Number of 5 card combinations} = \frac{52!}{5!(52-5)!} = 2{,}598{,}960$$

If all hands are random, a specific hand probability is approximately 1 out of 2,598,960 or about 4 in 10 million. Any specific hand is a very rare event.

As previously stated, a flush is a hand with all cards the same suit. There are exactly 4 suits, and so a flush contains 5 of the 13 cards in a suit. The number of different flushes follows closely the combination formula:

$$\text{Combinations within a given suit} = \frac{N!}{(r!)(N-r)!} = \frac{13!}{5!(13-5)!} = 1287$$

But there are four suits. Consequently, there are 1287×4 possible ways to get four cards all the same suit.

$$1287 \times 4 = 5148$$

The probability of being dealt a flush in a game of five-card straight is

5,184/2,598,960 or .002 or 2 chances out of a thousand

It is interesting to note that this is very close to the break point on a typical statistical control chart for an out-of-control condition. This will be covered further in Chapter seven, but there are many parallels between statistical process control, Six Sigma, the probability of rare events, and odds that show up in the often more familiar world of poker playing. When trying to explain the concepts of statistics to a sales manager or CEO (typically not familiar with the concepts of statistics), try explaining in terms of card playing or something more typical that they are used to in the real world. I have found poker hands to be more interesting and useful to illustrate the probability of an event than using examples such as the number of combinations of three candidates from five running for office.

Classroom activity 3.1: The poker hand flush by the "phone number" card trick

When fully practiced and utilized, the two card tricks that follow (which require minimum technical skill) are effective techniques to illustrate the concepts of probability. Since they are more interesting than typical classroom examples, they will also be easier for students to remember.

Step 1: Start by stacking a standard deck similar to examples in Chapter two

Start by stacking a standard 52-card deck in the following manner.

Take any flush combination and turn them facedown and then turn the rest of the deck face up. In a game of standard poker, a five-card flush is all cards dealt that are the same suit. Refer to Figures 3.1 and 3.2.

Place the stacked deck in the box. The next step is a trick designed to convince the students that they shuffled and cut the cards when actually they are only shuffling down to the last five cards in the deck.

Step 2: Placebo shuffle by students

Have three or four students randomly chosen for this exercise. It would be best if they are in front of you as if being dealt cards at a poker game.

Pull the cards from the case with the five flush cards facedown and the rest face up on top of the deck.

Deal out 6–10 cards to each person to shuffle. Make sure the bottom five cards (the flush) and, at a minimum, the one to five cards on top of the

Figure 3.1 The preload setup: the flush turned over at the bottom of the deck.

Figure 3.2 The deck reversed for viewing purposes.

five cards at the bottom are not passed out (Figure 3.3). Have the students shuffle the 6–10 cards handed to them.

After the first student to your left fully shuffles the first group, take those cards, turn them facedown, and place them at the bottom of the deck. It is critical that they are placed facedown, facing the same way as the hidden flush cards and opposite to the rest of the cards in your hand. Immediately, take the remaining cards in the deck down to the five flush cards (which will be the next cards facing down) and have the first student finish shuffling the last of the cards (Figures 3.4 and 3.5). Typical text

Figure 3.3 Dealing the first section.

Figures 3.4 and 3.5 The shuffled cards from the first person turned over and placed at the bottom and dealing out the remaining cards down to the flush to the first person.

at this point might be: "oh here first student, please take and shuffle the rest of the deck." This step is critical as it provides the image of the entire deck being shuffled when in fact the last five cards are not shuffled. This step only involves the first student. At this point, the flush is on top of the facedown cards, which gives the image the entire deck was shuffled, but it was not.

Take the remaining cards that have been shuffled by the other two to three students and place their cards at the bottom of the deck facedown (Figures 3.6 and 3.7). Now all cards in your hand are facing the same way. Take the last 6–10 cards from the student who has them and place them at the bottom as well—this student shuffled two sets of cards. This has the effect of taking all the cards except the prestacked flush, which is now on the top of the deck (Figure 3.8) and can be dealt out to the student (Figure 3.9).

Figures 3.6 and 3.7 Returning student-shuffled cards to the bottom of the deck.

Figures 3.8 and 3.9 The flush is now on top of the deck and in the student's hand when dealt.

Step 3: Determining probability of flush and revealing the flush

The rest of the trick is all demonstration skills. Typical dialogue used for this section is: "What are the odds of being dealt a flush? What are the odds of not getting a flush?"

Compare the odds to comparable real-life events. What are the odds of a typical highway accident? What are the odds of an out-of-control point on a control chart?

Some other comparable tricks in computations that can be used by this method include the following:

1. Stack the deck with a straight flush (a straight flush is all the same suit and the cards in order, such as 4-5-6-7-8-9 of clubs). What are the odds of a straight flush, and how would it be computed? The probability? .00001539. This one is relatively easy to compute

as there are only 10 possible straight flushes per suit and four suits for a total of 40 possible combinations. Utilizing the same five-card combinations for a deck of 52 as used earlier (2,598,960) provides the probability.

2. For a relatively easy but interesting demonstration, prestack the deck with 10 cards, alternating two five-card straight flushes (Figures 3.10 and 3.11). Follow the same procedure, but at the end take two students and deal out every other card to those two students. Each will have a straight flush. Walk through the odds of two straight flushes, and then have the two students turn over the cards to reveal the odds. The odds of this happening are less than most national lottery systems. This can be computed using the multiplication principle from Chapter two and from the odds of a single flush. Approximate odds: .000004 or very close to the odds of a defect at Six Sigma levels.

Figures 3.10 and 3.11 Stacking the double flush and final deal of the double straight flush.

Classroom activity 3.2: Phone number card trick to elaborate on the concept of combinations–permutations

Let's take a typical long number combination in which order is critical and use this to demonstrate how rare events are possibly computed.

Take a fully shuffled deck of 52 cards. Take a student's phone number—for this next section, I will use 564-203-1789. If each number is represented by a card number (the queen will represent the number "0." Look closely at the queen's letter q on all queens in the deck, and it looks just like a "0"), what are the odds that in a fully shuffled deck, the dealer will deal out your phone number? Interesting question—seems very remote.

Unlike the previous example with dealing a flush, the distinct order of the cards being revealed does matter. Consequently, the formula for combinations is modified and becomes what is called permutations:

$$\text{Number of 10 card combinations} = \frac{52!}{(52-10)!} = 5.7 \times 10^{16}$$

So, are the odds 1 out of 570 quadrillion? Close, but there are four suits for each card.

The number of 10-number phone number combinations in a deck of 52 cards by the combinations formula and modified for four cards per each suit is:

$$\frac{52!}{(52-10)!} = 5.7 \times 10^{16} \times 4 \text{ suits} = 2.2 \times 10^{15} \text{ or 1 out of 2.2 quadrillion!}$$

A rare event by any account!

Notice the difference between this and the first example with the flush. Being dealt a hand of 2 of hearts, 4 of hearts, 8 of hearts, 9 of hearts, and 5 of hearts is the same as 2 of hearts, 5 of hearts, 8 of hearts, 4 of hearts, and 9 of hearts. The phone number requires a distinct order to the cards. Being dealt a 10-card combination that could be rearranged into the correct phone number is a much different odds—but still very rare.

Step 1: The phone number card trick: Stacking the deck

Stack the deck with a student's or your phone number—this method will work with any unique set of numbers that does not have a repeated digit more than four times. Once again, for this example, we will use the phone number 564-203-1789 (Figure 3.12). For added effect, take the phone with

Figure 3.12 The bottom stack of the phone number.

the number and place it in a closed bag in the middle of the classroom or some unique location. The reason will be shown at the end.

Step 2: Fake shuffle the phone number from the bottom of the deck to the top

Repeat Steps 1 and 2 of Activity 3.1 (the flush trick) through Figure 3.8 stopping with the phone number on top of the facedown deck. The cards should appear as shown in Figures 3.13 and 3.14.

Add in the additional steps to give the image of not only shuffling the deck, but also cutting the deck.

Step 3: False cut of the deck

At the end of Step 2, the phone number should be the top 10 cards of the deck. Notice that if you want, just like the flush in the prior example, the top 10 cards can be dealt out in order. But to solidify the illusion that the cards are truly randomized, add the following steps. For these next steps, it is best to procure a deck of cards and perform each step.

Step 4: The first random cut

Have a student cut the deck removing approximately the top half, using the statement "it works best if you cut near the middle" (Figure 3.15). That statement is added to ensure the deck is not cut at the top 10 cards, which is the phone number.

Figures 3.13 and 3.14 Image of the deck shuffled when the top is the set up phone number.

Step 5

Take approximately half the deck from the student; return it to the deck flipped over on top, that is, now facing up (Figure 3.16). This is a critical step easily missed—the cut portion must be returned to the top of the deck facing up. The phone number will be facing up at the bottom of the half deck but just placed on the deck out of clear view of the student.

Step 6

Have the student cut the deck a second time. This time, use the statement, "take a really deep cut." The next cut must be below the top half of the deck between the top half and the bottom of the deck. That statement almost

Figures 3.15 and 3.16 Cut the deck, flip the top half over, and return it to the top of the deck.

guarantees the cut will be below the top half. If in the extremely rare incident the student takes a cut on the top half in the section cut in Steps 4 and 5, have a second deck available hidden, and restart the complete trick over. Again, this is an extremely unlikelihood if told to "take a deep cut."

Take the cards from the student, turn the cards over, and replace on top of the deck (Figures 3.17 and 3.18). This has the effect of two halves of the deck facing in opposite directions. The bottom three-fourths of the deck is facing down, and the top 10 cards in the bottom three-fourths of the deck are the phone number cards.

Step 7

At this point, the top 5–20 cards will all be facing up and the rest facing down. The top of the facedown cards is the phone number. Throw off the top cards facing up, and discard them away from you (Figure 3.19). They are not needed for the remainder of this illusion.

Figures 3.17 and 3.18 Cut the deck at the three-fourths point, flipped over, and returned to the top of the bottom.

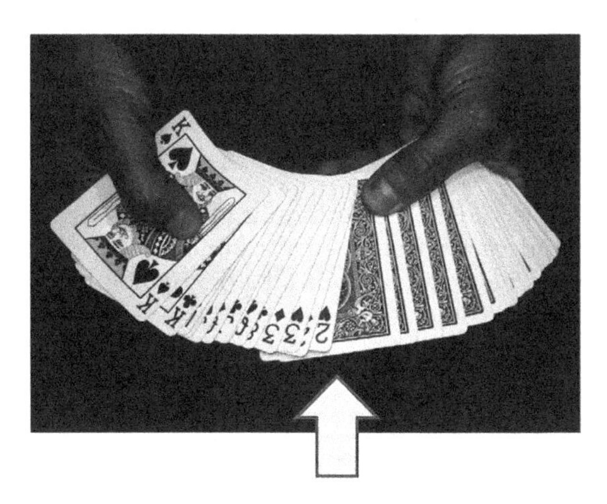

Figure 3.19 Remove top cards down to the first face card (arrow) and discard.

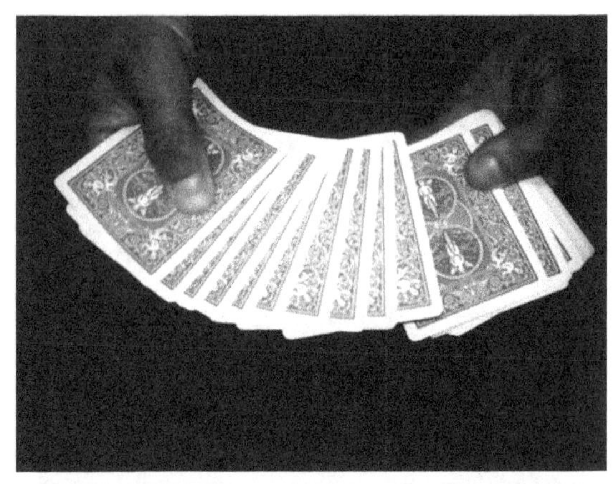

Figures 3.20 and 3.21 The phone number.

Count the top 10 cards from the top of the cards facing up (see arrow in Figure 3.19). After you have thrown off the cards facing up, these will be the top 10 cards and the phone number in order (Figures 3.20 and 3.21).

Step 8: Reveal the phone number

The rest is dialogue and can be customized to the specific situation.

Have one student read off the card numbers to you and transpose to a white board. The only issue with the dialogue might be to convince the students that ace is the number one and a queen is the number zero. Typical dialogue:

> What number does the ace best represent? This should prompt the number 1. As an alternative: how many spades, diamonds, clubs, or hearts (depending on the card chosen) appear on the card? That will be one.

For the queen. What number looks most like the queen? It will almost always be the number zero—look closely at the letter Q, and notice the resemblance. If there is any difficulty in this step, change the question to a leading question—does this card look like a zero more so than any other card?

Have someone dial the phone number. If it is in the room in the concealed location, the shock value will be significant. Answer the phone with "the odds of this happening are 1 in 100 trillion." Then work through the formulation on the board. The shock value will be heard throughout the building.

Special notes concerning this trick: The example phone number had 10 unique numbers. This illusion works up to four repeats, which is most phone numbers. If there are four nines in the phone number, use the four nines of the four different suits.

If for some reason this does not work or a card is returned to the wrong spot, have a backup deck hidden ready to start the trick over again. For those familiar with the concept behind Failure Mode and Effects Analysis (FMEA), a card returned to the wrong spot would be a very high level on the Risk Priority Number scale.

The phone number trick and "useful, relevant, and interesting/shocking?"

The previous two card tricks are designed primarily for one reason—to increase how interesting it was to learn about probability. Recall the theory of useful–relevant–interesting/shocking; the more the combined value of the three, the more it will be stored in memory. Using a shocking trick likely moves the combined index up significantly. Why? It is in the last of the three—interesting/shocking. Try this test—have the students compute the odds of a rare event using either combinations or multiplications 1 week later. Notice the recall. Try it again, 1 month later. The results will be surprising.

Other card tricks to be used with probability theory

- A random student's driver's license number
- The number π to seven significant digits (3.1415926)
- The phone number for the University Chancellor
- Three standard deviations to seven digits
- The address of the US White House.

A follow-up team exercise for students or in a continuous improvement class

This is a rare event that should not have happened. It did not happen by chance. In a real-life situation, this was a special cause of variation. In a manufacturing operation, this is a once-in-a-lifetime opportunity to significantly improve a process.

In work teams of 2–5, take the process from start to finish and draw a flowchart on a white board. For examples of how to make a flowchart, see the reference in the bibliography. Assign probability of failures at each step of the process. What are difficulties in recall? What are the difficulties in determining probabilities?

When the failure analysis is completed, reenact the card trick. Who can determine how it was done?

Author's notes

Properly executed, the phone number or poker hand combination works for memory retention as most of all my past students still remember this exercise. Several students have come back later and computed on their own other rare events in game theory, gambling, and so on. Replacing a standard academic exercise with this method will likely break the resistance to learning a new and different area.

Some cautions and recommendations from past classes:

- This exercise takes practice as it is easy to forget steps. Best to practice 50 times and a minimum of 10 without failure.
- This is an excellent time to get creative. If the phone number is utilized, it does not have to be your own. A concept used by magicians caused "forced randomization" that can be utilized to convince the audience that a certain student was picked by luck when in fact they were chosen out prior to class. Prearrange to have a relative or friend available at that time. Use that person's phone number for the exercise and have the student call that number. This is shocking as the student and participants in class do not believe they were part of it.
- This can and should open up an entire discussion on how to compute probabilities in other areas of interest. Probability in the sport of baseball is an excellent use of the multiplication principle and combinations. The students should be able to compute the odds of many other poker hands such as a full house, three of a kind, etc., as they are a natural continuation on the abovementioned methodology. Moreover, they are also generally more interesting than computing how many positions can be held by 3 candidates from 10 people.

Bibliography

Chen, Bill; Ankenman, Jerrod, *The Mathematics of Poker,* 14–16. 2006, ConJelCo LLC, Pittsburgh, PA.

Hayes, William L., *Statistics,* 5th Edition, 130–135. 1986, Holt Rinehart and Winston, Inc., New York.

Menenhall, William; Reinmuth, James E.; Beaver, Robert; Duhan, Dale, *Statistics for Management and Economics,* 5th Edition. 1986, PWS Press, Boston, MA.

chapter four

Bayesian statistics

When accurate past information shortens the distance to the aha moment

Objectives

- Review of the basics of Bayesian statistics.
- How to utilize Bayesian statistics in many practical applications.
- A simple coin trick to utilize in the explanation and teaching of Bayesian statistics.

At the start of the 2016 baseball season, the Kansas City Royals were the odds-on favorites to win the World Series. Just coming off of winning the World Series in 2015, they were the team to beat. But, something happened as the team went through the season. Their star player hit a batting slump. Injuries to key players mounted. A less than 0.500 winning percentage by mid-May began to lower the odds of them making the playoffs, let alone wining the World Series. As the season went along, the odds of them winning the World Series dropped significantly based on the new updated model.

A manager of a sporting goods store in Kansas City orders Kansas City Royals jerseys by the thousands based on his projection of them reaching the playoffs and winning the World Series. As the season slumps, his projection is updated, thus lowering the Kansas City Royals jersey orders.

The local civic theater manager orders concession food and programs for an upcoming show based on the preview of audiences from other cities. After the first night, the local critics give a not-so-glamorous review. The expected sales are reduced by 20% for concession sales and programs.

Welcome to the work of Bayesian statistics. In all three of these scenarios, probabilities are altered based on new information. The updated probabilities are referred to as prior probabilities, which will be altered by new information. Knowing how reliable the new information is improves the new projection, because many times, an estimate of reliability has to be made of the new information.

Bayesian statistics was primarily attributed to Reverend Thomas Bayes (1702–1761). By all accounts, he was a successful mathematician who used mathematics and the evidence around him to study the existence of God. Reportedly, he felt his work was too controversial for publication and therefore he never published. He died never really knowing how popular his theory would become. It was eventually used to change the outcome of WWII, is used in hundreds of court cases, and started a firestorm of controversy in a *Parade Magazine* article written by Marilyn Vos Savant on the now famous Monty Hall door problem. Users of the theory started the International Society of Bayesian Analysis that is still growing 200 years later.

What is Bayes' formula?

This will be shown by a very basic example from an article published in the *New England Journal of Medicine* in 1978. Suppose you are tested for a rare disease that occurs in the population at 1%. Also, the test result is 95% accurate. This means that if you have the disease, the probability of receiving a positive test result is 95%. If you do not have the disease, the probability of receiving a positive result is 5% (100%–95%, in this example). If you get a positive test result, how likely is it that you have the disease?

- 95%
- 85%
- 16%

Sixty doctors at four Harvard Medical School teaching colleges were asked the question. Only 11 answered it correctly. So what is the answer?

16%

When teaching a Six Sigma course and showing the example, a student asked me "why it was not 95%?"

We will walk through two ways of computing the correct answer of 16%, but first let's do it with Bayes' formula.

The confusing formula

$$\text{P(A given B)} = \frac{\text{P(B given A)} \times \text{P(A)}}{\text{P(B given A)} \times \text{P(A)} + \text{P(B given not A)} \times \text{P(not A)}}$$

Seems messy until you plug in the numbers.

Plug the numbers into the formula and work through it.

- P(A given B) = the heart of Bayes' formulation. What is the probability that I have the disease given the positive result?

- P(B given A) = the more typical posterior statement. Given you have the disease, what is the probability it will be found? In this case, it will be 95% or the accuracy of the test. Notice the difference in wording between P(A given B) and P(B given A). The first is the probability that I have the disease given the test results; this one is given that I have the disease, how likely will it be found. Two different statements.
- P(A) = the probability that I have the disease. That is only 1%. Notice, how critical this number is in Bayes' formula? The rarer the event, the more it affects the outcome.
- P(B given not A). This is the probability of being diagnosed with the disease, given you do not have it. For this example, we will use 5%, although this does not have to be the compliment of P(B given A).
- P(not A). This is the probability that you do not have the disease, or in this case, 100%–1%, or 99%.

Plugging in the numbers, it works out like this:

$$\text{P(that you have the rare disease given that you tested positive)} = .95 \times \frac{.01}{(.95 \times .01 + .05 \times .99)} = 16\%$$

Interesting? If you get diagnosed for a rare disease by a testing method that is not perfect, don't panic—you probably do not have it. So why is it not 95%? Notice Bayes' formula consists of two error components: How accurate the test is and how rare the disease is. Take the extreme case. A biologically born woman has a full lab bloodwork, and comes back with a positive prostate-specific antigen test for prostate cancer. Ready to go in for prostate surgery? Probably not as the P(A) is now zero and working through Bayes' formula it works out to zero.

So, the top line of Bayes' formula $P(B|A) \times P(A)$ is really just the combination of the accuracy of the test and the historical data of the problem or the reason I was being tested. The first part of the denominator is the repeat of the above P(B given A) × P(A) plus the second component P(B given not A) × P(not A). The second part of that denominator is the probability of being tested positive if I do not have the disease multiplied by the probability that I do not have the disease. Notice the greater the difference between the numbers, the lower the Bayes' formula or the probability that I have the disease. By formula, what lowers the outcome? The more inaccurate the testing, the rarer the disease. If both are poor (as in the earlier case), the likelihood is low, even if given a positive test. If it is a common disease with a very accurate test, the better probability that you have it (e.g., the common head cold). Also, notice P(A) or the probability that you have the disease, in this case 1%, changes with new information. If P(A) is 1% on the first test, it changes to 16% on the second test, when you take your positive test to a new doctor for a second opinion. New information changes Bayes' formula. This one can be changed based on subjectivity—but be careful. If the second test is positive and the test is for prostate cancer in a biologically born female, it still can't happen and zero is still zero. But if the lab calls up with a positive test for a biologically born male with breast cancer, the prior probability is that it is not zero. In an industrial setting, imagine a customer calls up and says the part spec is 10 mm ± 1 mm, but they just checked a part and it was 8 mm. You say that we have never had it happen that low, the probability is zero, and that it must have been someone else's part. Correct or not? Probably not true, unless there was perfect evidence that the part could not physically have been produced at that level—i.e., beyond the means of the die dimensions. Notice how this one can be subjective and many times be wrong. Gut feel never trumps evidence as we will see in a following coin trick example.

How does it work? Back to the prior problem. The first outcome is a 16% chance that I have the disease. What if a second test is taken to the same lab (notice that does not change the 95% test accuracy) and the results come out the same? Seems like a sure shot now that I have the disease—using the multiplication formula from Chapter two and assuming the lab accuracy did not change; should it be 1 – (.05 × .05) or 99.75%?

Returning to Bayes' formula for the second test and the four critical calculations making up Bayes' formula:

- P(B|A) = .95. Unless the lab changed or they changed their instruments, this one is constant at 95%.
- P(A) = .16. Notice this was 1% based on historical information, but changes with new information. After the first test was positive, we now utilize the prior outcome from the first Bayes' formula. In other words, we now expect going into the second test that the patient has a 16% chance of having the disease.
- P(B given not A). This one does not change. This is still related to the accuracy of the test and in this case is still 5%.
- P(not A). This part changes with new information. For the second test, it is 84% (complement of 16%).

Working through Bayes' formula on the earlier information:

$$\text{P(A given B)} = \text{P(B given A)} \times \frac{\text{P(A)}}{\text{P(B given A)} \times \text{P(A)} + \text{P(B given not A)} \times \text{P(not A)}}$$

$$P(\text{A given B}) = \frac{(.95 \times .16)}{(.95 \times .16 + .05 \times .84)} = 78\%$$

Wait a minute. Even after the second positive test result, there is an almost one in four chance that I still do not have the disease. Still correct, as it was a rare disease and the test was not very accurate. One more test to prove beyond reasonable doubt. The third test by Bayes' formula, now using 78% as our prior probability:

$$P(\text{A given B}) = \frac{(.95 \times .78)}{(.95 \times .78 + .05 \times .22)} = 98\%$$

Still there is a 2% chance of being totally wrong even after the third test. Only after the fourth positive result, does it start to approach reasonable odds that I have the disease (Table 4.1).

We discussed that having a low accuracy test, as well as testing for a rare disease, contributed to a relatively low chance of actually having the disease, given a positive test. How will those numbers chance with a more accurate test, or a more prevalent disease?

First, what happens if the instrument was recognized poor at best and was replaced? What changes in Bayes' formula if the instrument (and accuracy) was changed? (Table 4.2)

Table 4.1 Summary of test results with gauge accuracy of 95%

	Test result with accuracy of 95%	Probability that I have it (%)
First test	Positive	18
Second test	Positive	78
Third test	Positive	98
Fourth test	Positive	99.9

Table 4.2 Summary of test results with gauge accuracy of 98%

	Test result with gauge at 98%	Probability that I have it (%)
First test	Positive	33
Second test	Positive	96
Third test	Positive	99.9
Fourth test	Positive	99.9999

Table 4.3 Summary of test results with gauge accuracy 95% and prior history of disease

	History of disease (%)	Probability that I have it (%)
First test	1	16
First test	5	50
First test	20	82

Second, what happens if the prior disease was not so rare? Let's take the same poor instrument and change the prior information. Say our best guess is that the disease happens not at 1% but at 5% or 20%. How does that change the outcome of the first test? (Table 4.3)

Classroom activity 4.1: Coin trick to explain Bayes' theorem and illustrate the problem with subjectivity in prior knowledge

This is an excellent example of the use of an easy coin trick to explain Bayesian statistics and the problem with subjectivity or opinions entering into prior knowledge.

Procure a two-headed quarter for this exercise.*

Step 1: Setup and identifying the two-headed quarter

First is the setup by the instructor. The weight of the two-headed quarter is different from a normal quarter. With about 2 min of practice, the weight of the quarter can be detected in the instructor's hand. Being able to detect the difference is critical to this next step but very easy to learn (Figure 4.1).

Step 2: Pass out quarters and flip three times

Take three quarters out, one of which is the fake two-headed quarter. Mix up the coins in a jar or any item giving the impression that the coins are totally mixed up. Pull out each quarter individually from the jar passing them to three students—one of which will be the fake quarter, and you and only you will know which one has the fake quarter. Inform the three students that one of the coins is fake and that the other two are real. Make sure they do not inspect the coin! They do not know which one is

* To purchase two-headed coins, visit: http://www.twoheadedquarter.net/.

Figure 4.1 Two-headed coin.

fake just that one of them is two-headed. Only you know who has it by the weight. Ask the one with the fake coin to flip the coin three times—which will come up all heads. The question to ask the students is this: Given your knowledge of the instructor, what is the probability that the coin just tossed, which came up heads three times, is the fake two-headed coin? This is a classic Bayesian statistics problem as the outcome to the question depends on the strength of our prior belief that the instructor knows which coin is the fake coin and which coin the instructor chooses.

It all depends on your prior belief that the instructor chose the coin randomly, chose the fake coin, or chose the real coin. Let's work through each scenario separately.

Scenario 1. I don't think the instructor had prior knowledge; the coin was randomly chosen

Working through the formula:

$$\text{P(A given B)} = \text{P(B given A)} \times \frac{\text{P(A)}}{(\text{P(B given A)} \times \text{P(A)} + \text{P(B given not A)} \times \text{P(not A)})}$$

- P(B given A) = probability that I will get three heads given it is the fake coin. That is equal to 100% or 1 as it has to come up with three heads.
- P(coin was fake) = this one changes with your belief that the instructor knows which coin is fake. In this scenario, we do not think the instructor had prior knowledge—in this case, it is 1/3, as the coin was randomly chosen from three possibilities.
- P(B given not A) = this is the odds by the multiplication formula that if the coin was not fake but real, then we could get three heads. This is $1/2 \times 1/2 \times 1/2 = 1/8$.

- P(not A) = this is the probability that it is a real coin and not the fake coin, or 2/3.

P(fake coin given three heads) =

$$\frac{\text{P(three heads given fake coin)} \times \text{P(fake coin)}}{\text{P(three heads given fake coin)} \times \text{P(fake coin)} + \text{P(three head given not the fake coin)} \times \text{P(not the fake coin)}}$$

$$\text{P(fake coin given three heads)} = \frac{(1 \times 1/3)}{((1 \times 1/3) + 1/8 \times 2/3)} = 4/5 \text{ or } 80\%$$

Scenario 2. I think the instructor had prior knowledge and chose the person with the fake coin

$$\text{P(A given B)} = \frac{\text{P(B given A)} \times \text{P(A)}}{(\text{P(B given A)} \times \text{P(A)} + \text{P(B given not A)} \times \text{P(not A)})}$$

- P(B given A) = probability that I will get three heads given it is the fake coin. This one does not change based on my prior knowledge.
- P(coin was fake) = this one changes with your belief that the instructor knows which coin is fake. In this scenario, we believe very strongly that the instructor knows which coin and chose the person with the fake coin. This is the extreme case and is 100%, but based on the posterior knowledge of getting three heads.
- P(B given not A) = this is the odds by the multiplication formula that if the coin was not fake but real, then we could get three heads. This is $1/2 \times 1/2 \times 1/2 = 1/8$ and did not change.
- P(not A) = this is the probability that it is a real coin and not the fake coin. In the extreme case, this is zero based on our opinion that the instructor knows which coin is fake.

$$\text{P(fake coin given three heads)} = \frac{(1 \times 1)}{((1 \times 1) + 1/8 \times 0)} = 1 \text{ or } 100\%$$

Scenario 3. I think the instructor had prior knowledge and chose one of the real coins

$$\text{P(A given B)} = \frac{\text{P(B given A)} \times \text{P(A)}}{(\text{P(B given A)} \times \text{P(A)} + \text{P(B given not A)} \times \text{P(not A)})}$$

- P(B given A) = probability that I will get three heads given it is the fake coin. This one does not change based on my prior knowledge.

- P(coin was fake) = this one changes with your belief that the instructor knows which coin is fake. In this scenario, we believe very strongly that the instructor knows which coin and chose the person with the real coin. This is the extreme case and is 0%.
- P(B given not A) = this is the odds by the multiplication formula that if the coin was not fake but real, then we could get three heads. This is $1/2 \times 1/2 \times 1/2 = 1/8$ and did not change.
- P(not A) = this is the probability that it is a real coin and not the fake coin. In the extreme case, this is 100% based on our opinion that the instructor knows which coin is fake.

P(fake coin given three heads) =

$$\frac{\text{P(three heads given fake coin)} \times \text{P(fake coin)}}{\text{P(three heads given fake coin)} \times \text{P(fake coin)} + \text{P(three heads/not the fake coin)} \times \text{P(not the fake coin)}}$$

$$\text{P(fake coin given three heads)} = \frac{(1 \times 0)}{((1 \times 0) + 1/8 \times 1)} = 0\%$$

Notice scenarios 2 and 3, the outcome will depend on how strong the subjective belief that the student has in that the choice was completely random or their belief that the instructor can tell from the jar that the coin was fake or not.

This exercise works excellently at understanding Bayesian statistics as it puts the formula into a real exercise—and that is not common.

The medical problem by tree diagram

It is somewhat difficult to understand Bayesian statistics by Reverend Bayes' formula. Let's prove that it works, by another method—a tree diagram. See the example in Figure 4.2. In this example, we take 10,000 patients and work through the probabilities.

Utilize the Bayes' formula and follow through on the tree diagram.

$$\text{P(B given A)} = \frac{\text{P(A given B)} \times \text{P(A)}}{(\text{P(A given B)} \times \text{P(A)} + \text{P(A given not B)} \times \text{P(not B)})}$$

What is that term in the numerator that is duplicated in the first half of the denominator? That is the probability that I do have the disease and I am correctly diagnosed with it. Or in our 10,000 patients, that only happens 95 times. So what is the other term in the formula (the second half of the denominator)? That would be the probability of being diagnosed with

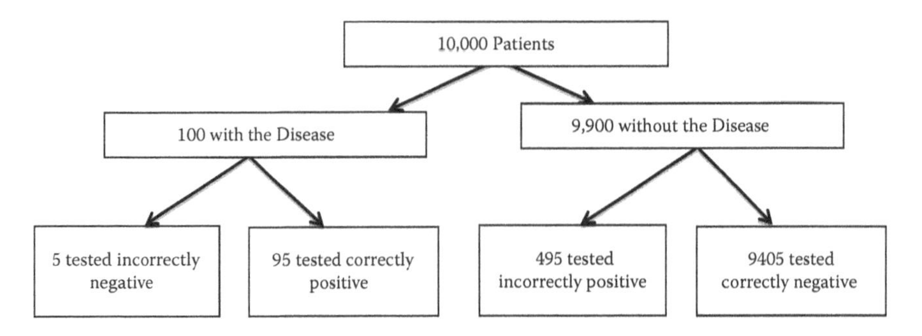

Figure 4.2 Tree diagram of disease at 1% of population and gauge at 95% accurate.

the disease when I do not have it. In that case, there are only 495 incidences out of our 10,000 patients. Go through the math, and the results look like this:

$$\frac{95}{(95+495)} = \text{approximately } 16\%$$

It will likely help in the application of Bayes' formula to work through an actual application using tree diagrams.

Your honor, gut instinct does not counter Reverend Bayes

Bayes' formula is used often in court cases to counter a jury's intuition. Take the very famous case of Sally Clark in 1999. She was brought to trial and convicted for double homicide of her two infant sons. Her defense was not murder but rather sudden infant death syndrome (SIDS). Two competing theories. Murder or SIDS death. Prosecutors said it was murder. Bayes to the defense.

The suggested cause of death by the defense was for SIDS, which was incorrectly originally determined to be 1 in 73 million and later revised to 1 in 130,000 based on the multiplication principle, from Chapter three. It was incorrectly calculated by assuming SIDS deaths were totally independent events, which was later proved not true. The probability of a second sibling dying from SIDS increases significantly if the first child died from SIDS. The concept of double SIDS death seems very rare unless put into the concept of the likelihood of a mother killing both her children. But how do you compute this? Reverend Bayes to the rescue.

Is there a way to compute the odds of a mother killing her two young children? Of course, there is. Court records showed 650,000 births and

30 known and undisputed murders by mothers per year over a course of 5 years. Notice the language—accuracy in Bayes' formula is critical—it cannot be gut feel I think they were murders. These were 30 known infant murders per year in England and Wales over the 5-year time span. But that is for a single murder and must be modified for double murder. This is modified by a very conservative approximation of 1/10 as it is difficult to estimate this exactly. But if the single murders is .000046, we will estimate it to .0000046, which is likely an overestimate of the probability of a double murder—i.e., way in favor of the prosecution.

Inserting the numbers into Bayes' formula:

$$\text{P(SIDS given the data)} = \frac{1 \times .00000077}{1 \times .0000077 + .999992 \times .0000046} = .6$$

There was a bigger chance of dying of SIDS than of a double murder by the data and definitely not enough to convict beyond reasonable doubt. Of course, other evidence may influence the probability of double murder such as murder weapons, past history of the mother, etc., but without further evidence, there is more evidence to support SIDS than murder based on the historical data, which, in this case, comes into play.

Sally Clark spent 4 years in prison for the aforementioned crime before her conviction was overturned. The original evidence was presented as two independent events resulting incorrectly as the probability of 1 out of 73 million for the probability of two. Sally tragically died in 2007 of acute alcoholism never getting over being wrongly convicted. Likely as a result of her case, Bayesian statistics is becoming more acceptable in a court of law and was used in two follow-up cases in England to reverse convictions for SIDS deaths.

How it should work in industry (but rarely does)

Let's say an established machining process is running excellently with a fully mistake-proofed operation, and a control chart on the operation at a 5 sigma level (CPk of 1.67). Think of mistake proofing as any mechanism designed such that the defect cannot be physically produced or has a very low probability of being produced. The probability of failure is about .03% with the 1.5% controversial process shift. The process instrumentation is 95% accurate. The customer of the process randomly takes a part, measures it, and calls it out of specification. They call the supplier and demand an immediate sort of their facility.

What is the probability of the process producing a part out of spec given the part was measured out of specification? Note the language—the common answer is only 5%, but that is not taken into consideration that the part was being produced at .03 defect rate and the instrument was not accurate. Once again, Reverend Bayes to the rescue.

$$P(\text{A given B}) = \frac{P(\text{B given A}) \times P(\text{A})}{P(\text{B given A}) \times P(\text{A}) + P(\text{B given not A}) \times P(\text{not A})}$$

- P(A given B) = this is the heart of Bayes' formulation. What is the probability the part is bad given the bad result.
- P(B given A) = the more typical posterior statement. Given the part is bad, what is the probability that the part will be measured wrong. In this case, it is 95%.
- P(A) = the probability the part was made out of specification or in this case if the process is at 5 sigma levels is only .03%.
- P(B given not A) = given the part was good, what was the probability of being found bad. In this case, it was 100%–95%, or 5%.
- P(not A) = this is the probability the part was made incorrectly or 99.97%.

Working through Bayes' formula:

$$P(\text{A given B}) = \frac{.0003 \times .95}{(.0003 \times .95 + .9997 \times .05)} = \text{approximately } 6\%$$

Hold off the sort service. It very well may be another supplier's part or other highly unlikely events. But the chances of this part being out of spec are not 95% as many would argue, and resources spent chasing corrective action on this part can and should be used in other areas. Following the same path as earlier, there should be a second or third sample taken immediately, and then and only then can it be called out of specification. Arguably difficult to sell but normally the correct action.

Bayesian statistics is a fascinating area but defies common sense—as was found by the prosecutors of Sally Clark and found by the instructors at Harvard medical teaching college. But, it is a much underutilized tool outside of universities and statistics professionals. The technique was arguably used to shorten the length of WWII (as told by the 2014 movie about Alan Turing, *The Imitation Game*), helped start the insurance industry in the US, and demonstrated smoking causes lung cancer (believe it or not, at one time experts did not believe there was a connection).

Bayes' theorem is a very powerful but underutilized tool in industry, but is it that way because the theory is kept to the universities and PhD students? What would happen if this method was understood by all engineers, managers, nurses, and even manufacturing line personnel? Could a concept critical to understanding human life expectancy be used in the development of operator-less cars or by a line operator to find the solution to a perceived color difference in a cabinet-making operation? Reverend Bayes tried to use it to prove the existence of God. Although it may or may not have proved this successfully, it opened the doors for the prevention of millions of lost lives in WWII and saved Sally Clark from a lifetime jail sentence.

Follow-up additional classroom exercises:

- It can also be even expanded to the opening card trick in Chapter two of the aces and kings. Change the question to "what is the probability the deck was fully shuffled based on the new evidence?" Notice this is now a Bayesian statistics problem with prior and posterior distributions that change with every card that is turned over and the belief in the dealer.
- For the coin trick, change the number of coins passed out from three to four, but still only leave one as the fake coin. What changes take place in the analysis? Cover the same three scenarios of the three tosses. Change the number of coins to 10. What changes?
- Define a real-world scenario that applies to Bayesian statistics. Example: I have lost my keys in the house. Where do I look? Would Bayesian statistics suggest a random start to the search or a narrowed search based on common areas that they would have been placed?

Bibliography

Hooper, William, A closer look at Bayes' theorem and how it can defy instinct, *Quality Progress*, March, 18–22, 2014.

McGrayne, Sharon Bertsch, *The Theory that Would Not Die: How Bayes' Rule Cracked the Enigma Code, Hunted Down Russian Submarines and Emerged Triumphant from Two Centuries of Controversy*. 2011, Yale University Press, New Haven, CT.

Olofsson, Peter, *Probabilities: The Little Numbers that Rule Our Lives*. 2007, John Wiley & Sons, Hoboken, NJ.

part two

Introduction: Data, statistics, and continuous improvement via the sport of juggling

chapter five

Learning the sport of juggling step by step[*]

Objectives

After completion of this section:

- All instructors and students should, within 1 h, master juggling two balls or juggling scarves for an average of 10 tosses and begin to juggle 3 balls or 3 scarves.
- Understand the health benefits to the brain of juggling.
- Begin to understand the parallels between statistics and juggling.

The history of juggling and famous jugglers over the years

Juggling has a very long history. The picture in Figure 5.1 was taken of the Tombs of Beni Hasan estimated to have been constructed in 1994 BC. It was believed that this image is of a traveling group of entertainers. Three things are very prevalent in this picture—the subjects are all women; the third one from the left is believed to be performing a routine called "Mills' Mess," which is a very difficult maneuver that current jugglers give credit to juggler Steve Mills for starting in 1975. And finally, it appears that the second and fourth may have been passing balls. All three observations suggest the advanced state of juggling as an art and entertainment. All three suggest an advanced level of teamwork.

Let's move ahead to modern times to athletes and celebrities who have taken up juggling.

Many athletes have given credit for their hand–eye coordination to the sport of juggling. Following is a partial list of celebrities who are or were earlier active jugglers (list courtesy of Juggler Historian, David Cain):

Penn Jillette
Ronald Reagan

[*] To purchase juggling equipment utilized in this chapter, go to www.toddsmith.com, www. dube.com, www.higginsbrothers.com.

Figure 5.1 Image taken from the Tomb of Beni Hasan. (Courtesy of International Jugglers' Association.)

Kristen Stewart
Mel Gibson
Sarah Michelle Geller
Alan Alda
Dave Grohl
Paul Simon
Art Garfunkel
David Letterman
Christopher Reeve
John Denver
Brendon Fraser
Ben Stiller
Jackie Chan
Three Stooges
Charlie Chaplin
Buster Keaton
Marx Brothers
Harry Houdini
Will Wheaton
Jimmy Buffett
Richard Gere
Pamela Reed

Many politicians and actors are jugglers. Figure 5.2 shows the 40th President of the United States, Ronald Reagan.

And finally, generally regarded as the inventor of binary code and according to Bill Gates, "the father of the modern computer," is Claude

Figure 5.2 Ronald Reagan entertaining in 1939. (Courtesy of International Jugglers' Association.)

Shannon. Claude Shannon was probably more interested in juggling at MIT than noteworthy areas such as the development of binary code and the first computers. In his later years, Claude actually built a juggling robot and developed the juggling formula, which can be used to predict the number of balls a human can juggle based on input factors such as the height of the throw, the time in each hand, and the time each ball spends in the air. An interesting reference on Claude's formula development, the juggling robot, and his contribution to the modern computer can be found in the Bibliography. Claude was likely the first to develop the statistics behind juggling with the juggling formula:

$$N = \frac{H \times (F + D)}{(V + D)}$$

where N is the number of balls a person or robot can keep in the air and H is the number of hands (this is would be two unless two people are juggling, which would double this number and double the number of balls), D is the time in the hand, V is the time the hand is empty, and F is the time the ball is in flight. This model he used to develop the juggling robot was likely used to spark the creativity for his other world-changing invention.

Juggling as brain food

The use of juggling for entertainment has been well documented over the years, but recently, several studies have found a connection between the brain development, disease, and juggling.

A 2004 study for connection between the brain and juggling found a significant increase in the brain's gray matter in a group of 12 nonjugglers who were taught juggling and then juggled for a set amount of time every day. At the end of 15 days, the resulting MRI showed significant increases in the average frontal gray matter in comparison to a controlled group of nonjugglers. Details of this study are given in the Bibliography.

At the end of the study period, those 12 subjects quit juggling. After 15 more days, their gray matter actually decreased, approaching the starting level. The study concluded the daily practice of juggling could potentially reverse or delay the negative impact from Alzheimer's disease or even Lou Gehrig's disease.

With juggling, you never can reach a limit. Once better, you just add on another ball to the equation. That is not unlike most process development endeavors—once a problem is solved, it is on to more difficult problems. Juggling has been called the ultimate challenge of controlling patterns in space. In the forthcoming chapters, we will show how to use statistics and data to control those patterns.

The first step: The one-ball cascade

Let's start with one ball. For the remainder of this chapter, we will assume the person learning is right-handed. For those who are left-handed, reverse the hand notation as it is being described.

For this exercise, procure three round objects. As a guideline, tennis balls are too lightweight but the right size in diameter. Soccer balls are too big and golf balls are too small. A standard size orange works great, but be forewarned, after numerous drops (and there will be drops!), they have a tendency to crack open. Practice in a place where drops will not result in a long-term mess! Beanbags are excellent for learning to juggle as they do not bounce. Baseballs are also a good size, but they do roll around when dropped—but that may add to the exercise portion of the practice session!

Visualize a picture frame that you hold in front of you in your hands, and the top corners of the frame are about 15–25 cm above your head. Those will be critical targets for two-ball and three-ball juggling (Figure 5.3).

Figure 5.3 Visualize holding a picture frame shoulder wide with corners at 15–25 cm above head level.

Start with one ball, or for those who prefer to start with a slower pattern, use approximately 30 cm by 30 cm silk scarves. Silk scarves are an effective tool for learning juggling up to and including three and four objects, as the pattern is much slower.

Take the one ball, starting with arms bent and hands at waist level. Toss the one ball starting with your dominant hand (right hand for this analysis), tossing across the frame diagonally to the top left corner of the picture frame. Catch the toss with the left hand. Repeat the steps for the toss from the left side to the top right corner of the picture frame catching with the right hand at waist level. Do this exercise until 10 tosses can be made without dropping (number of tosses to drop will be critical measurement in the next four chapters). Critical items necessary to prevent dropping in this phase are as follows:

- Keep the eyes at the top corner of the picture frame. Practice not looking at your hands. This will be difficult at first but becomes more natural after about 5 min of tosses. This technique is critical to minimize the movement of the head, which results in the movement of the body. Movement of the body adds excessive variation to the entire process, causing the body to "overcompensate" for the movement.

- Keep the toss in a vertical plane in front of you at the natural length of the arms from the elbow to the hand. There will be a natural tendency for the body to want the balls tossed away from the body. This is a natural body protection mechanism to prevent being hit by one of the balls. If this problem persists, practice about 60 cm away from a wall.
- This will be a slow pattern. Count to yourself as you do this to establish the beat. The beat will be used when more balls are added.
- Record the hand that drops. Normally one hand is stronger than the other. This will change over time as your nonnatural hands for tossing and catching become accustomed to the process. One of the strongest benefits of juggling is the development of your left body side if you are right-handed or right side if you are left-handed, Cincinnati Bengals All Pro wide receiver, A.J. Green reportedly does not have a dominant hand—he is equally coordinated in both. He attributes this to his skill in juggling.
- Practice not moving your feet. The feet stay in one spot for this pattern and all other patterns except for a few advanced maneuvers.
- Alternate starting from the left and right hand. This will work to strengthen the weakest side.

Keep this pattern going until a consistent pattern is reached for an average of 10 tosses. This should last only about 5–10 min until the 10 tosses are reached (Figures 5.4 through 5.6).

Two-ball cascade juggling: The most critical pattern to learn

The two-ball cascade is done only after mastering the one-ball cascade, as learning this step is the hardest part of learning the three-ball cascade. For most people, if they master the two-ball cascade, the three-ball cascade can be learned relatively quickly. There are commonalities of the two-ball cascade with the three-ball cascade. It is the same pattern from left side to right side and reversed—left side to right side as the one-ball cascade. The arms and hands are held at the same positions—bent and parallel to the ground. The toss is still to the same corners of the imaginary picture frame. The eyes never look at the hands—look at the corners. This reduces the head movement, which in turn reduces the body movement, which in turn reduces the variation requiring the body to adjust the process.

Start the two-ball cascade with one ball in each hand. Toss from the right hand to the upper left corner of the picture frame. As soon as the ball reaches the peak, which should be the imaginary target, throw the left hand to the right upper corner of the frame. As soon as the toss is made

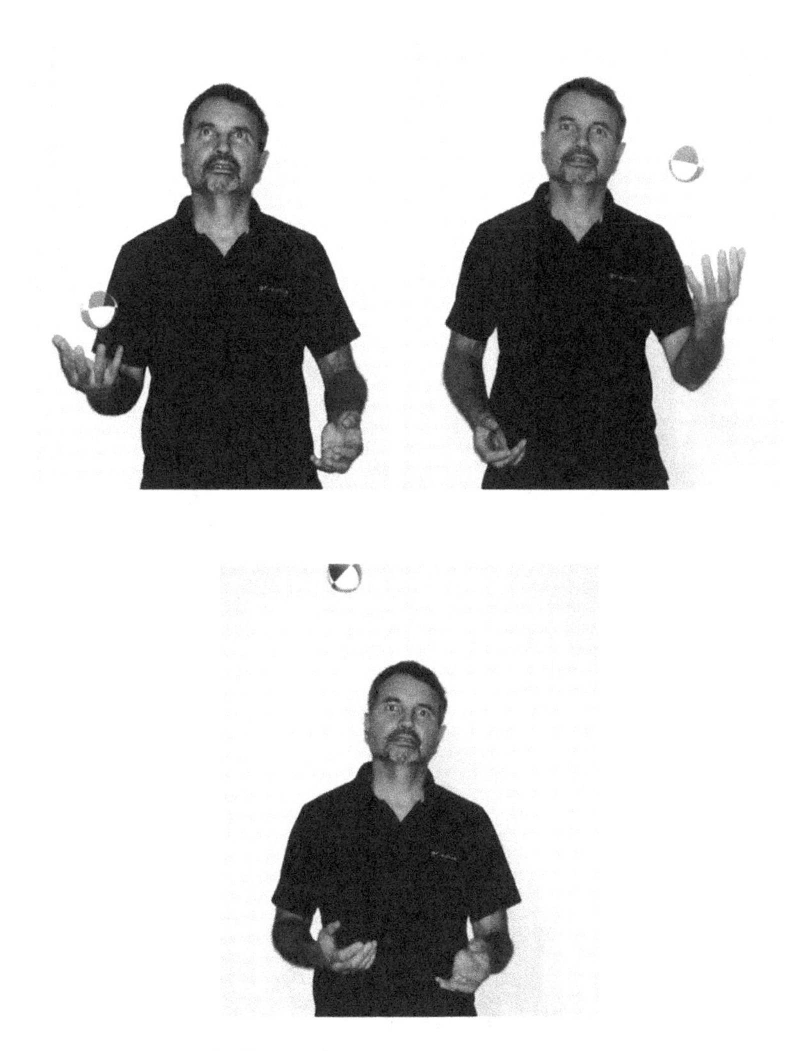

Figures 5.4–5.6 One-ball cascade.

with the left, the left hand catches the throw from the right side. As soon as the toss from the left side reaches the right side of the frame, the next toss is made by the right hand, and the right hand catches the toss from the left hand. This pattern is repeated until a ball is dropped (Figures 5.7 through 5.9).

Count the number of tosses until drop. This will be used as a gauge for continuous improvement in the following four sessions. Keep the pattern going until drop—do not stop with both balls in each hand

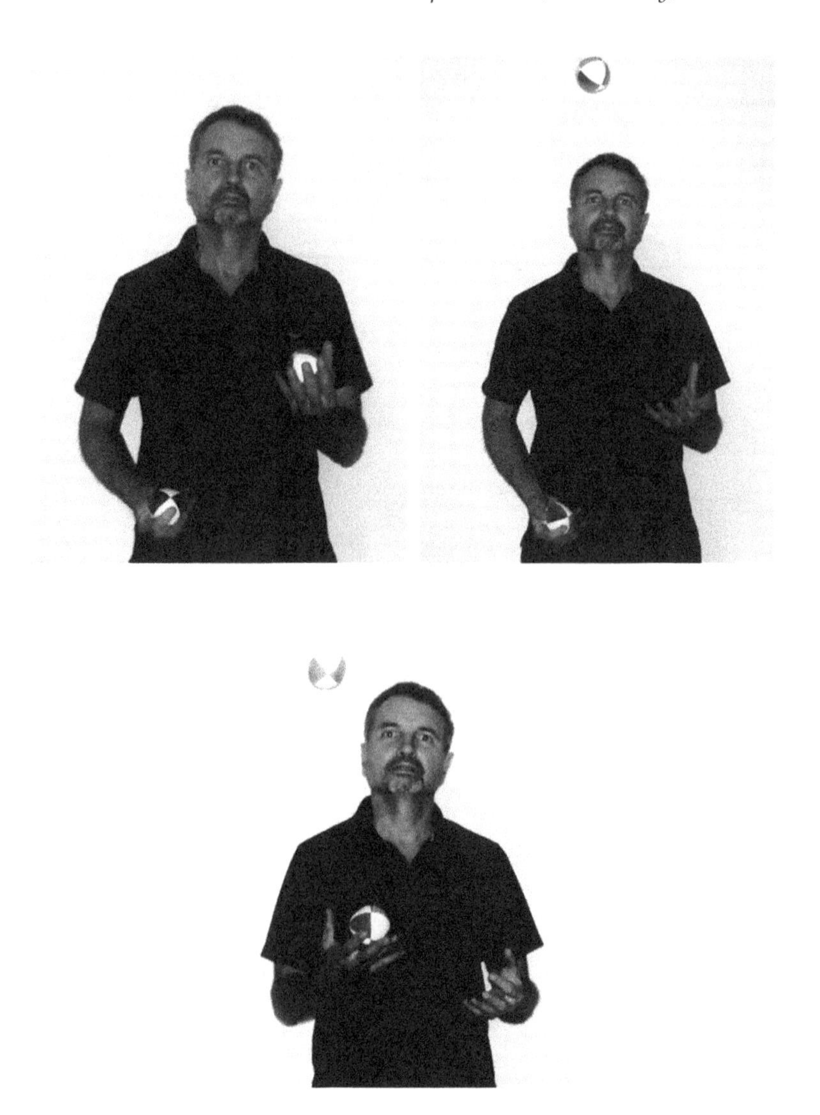

Figures 5.7–5.9 Two-ball cascade.

until mastering this pattern. Mastering this pattern is when 10–15 tosses without a drop are achieved.

There are common problems associated with this pattern. Most can be overcome with practice, but in some situations, breaking the cycle requires a change in the process (not unlike any manufacturing or nonmanufacturing process). Here are a few of the common problems and suggested solutions:

- Instead of tossing a ball to the corner of the picture frame, the ball is handed from one hand to the other. This is a common problem with new jugglers, and the habit must be broken. To break the habit, it is sometimes helpful to learn why this happens. The brain is actually "panicking" and rushing the process. Like many processes, attempting to operate beyond the natural speed results in off quality and downtime. The key to breaking this pattern is to slow down and make sure you are looking at the corners of the frame and not the hands. Record which hand is handing off and not tossing. It normally is the off hand of the juggler—for a right-handed juggler, it is the toss from the left side to the right side and for a left-handed juggler, the toss from the right side to the left. If slowing down and looking at the corners does not work, go back to the prior one-ball cascade and practice on the side that is the problem until the pattern is broken, or practice with scarves, which almost always breaks the pattern.
- The pattern moves outward away from you requiring you to "chase" after the pattern. Similar to the one-ball cascade, the body is tossing the balls away to prevent from getting hit by a ball. Like the one-ball cascade, move in front of a wall. This will prevent the body/mind from tossing the balls away from you.
- The pattern is stopped after a minimum number of throws with balls in both hands. This requires breaking down the process to microsteps. Which hand is stopping the process? Left or right hand? After determining the hand that is stopping, return to the one-ball cascade and repeat with that hand for a significant amount of time. When the pattern is broken, return to the practice exercise.

Continue the troubleshooting until obtaining 10 throws without dropping. As a qualitative measurement, add in the number of quality throws to drop. What constitutes a quality throw? Any throw in which the feet do not move and no ball is handed from one hand to another. The latter will likely result in a drop, however. The former will likely result in a drop

Table 5.1 Example format

Successful tosses before drop	Quality tosses caught without moving feet or throw from hand to hand	Opportunities per throw (throw on target and catch)	Maximum quality opportunities between drops
5	4	2	4
20	15	2	10
50	48	2	30
80	65	2	65

within the next one to three throws. Nonetheless, this technique will be a close analogy to the conventional Six Sigma metric, quality parts per opportunity, or quality parts between drops (Table 5.1).

Next step to three-ball juggling: The two-and-half-ball cascade

The pattern on this one is the same as the two-ball pattern with one exception: The right hand will hold the third ball in the palm of the hand (see below Figure 5.10 through 5.14).

Figures 5.10–5.14 Two-ball juggling with third held in right hand.

(Continued)

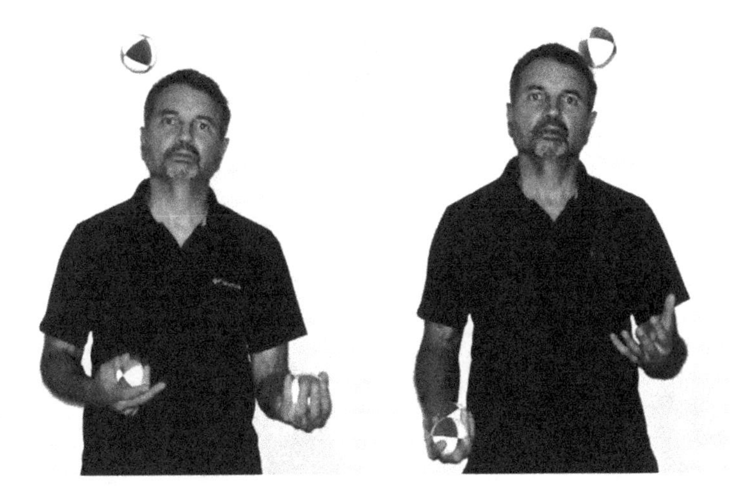

***Figures 5.10–5.14* (Continued)** Two-ball juggling with third held in right hand.

For reasons that are not entirely clear to avid jugglers, for most students of the sport, this will naturally add a delay in the two-ball pattern—essentially slowing the two-ball cascade learned earlier and adding in a spot in the pattern for the third ball—i.e., the source of this step's name, the two-and-half-ball cascade. Continue with this step until 10 consecutive throws are completed and repeat for the opposite side, holding the third ball in the left hand. This will create the three-ball pattern in two balls with the third ball being held in either hand, significantly reducing the learning curve for the three-ball cascade.

The three-ball cascade

Start with two balls in your dominant hand and one in your off hand. To start the pattern, the hand with the two balls throws the front ball to the targets practiced in the two-ball or two-and-half-ball pattern. The targets are exactly the same—the top corners of the imaginary picture frame. In the three-ball cascade, the pause that was likely built in under the two-and-half-ball pattern is now filled by the third ball. Although it may appear as though the pattern is faster, the pattern is exactly the same as the two-and-half-ball pattern or the two-ball pattern except the pause in those patterns is now filled with the third ball. See Figure 5.15 through 5.17 for pictures of the three-ball cascade.

Just like the two-ball pattern, the issues associated with the three-ball pattern are common to most first-time jugglers and can be corrected as follows:

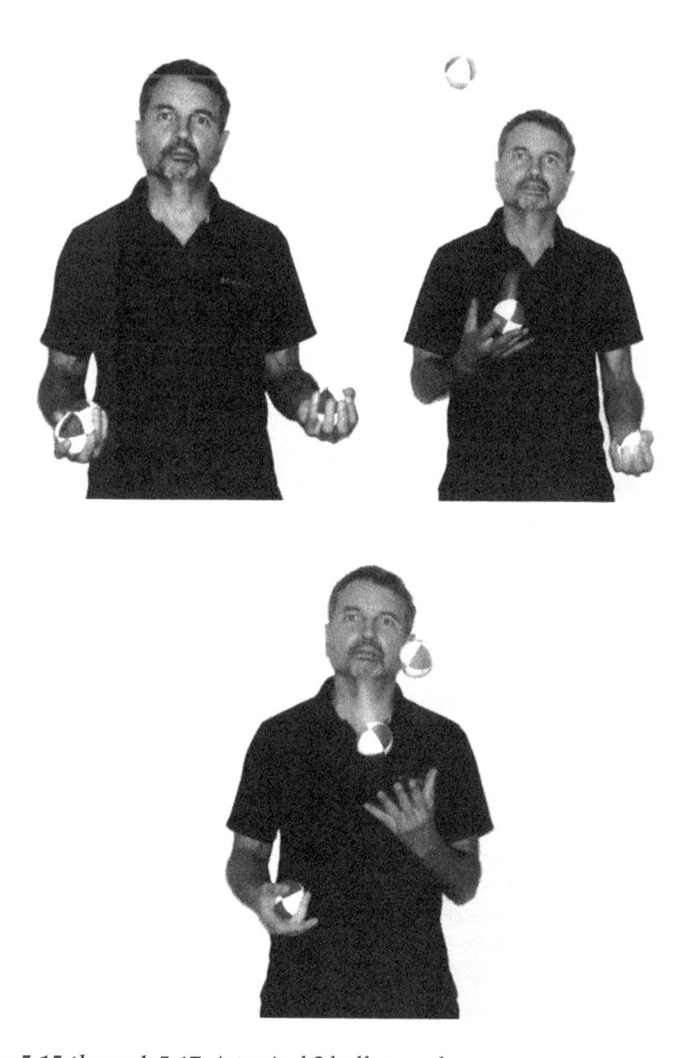

Figures 5.15 through 5.17 A typical 3 ball cascade.

- The pattern keeps moving out from the imaginary plane in front of the body, requiring you to lunge for or walk to the next ball. This is the same as with the two-ball pattern and results from the body's natural tendency to prevent it from being hit by one or more of the balls. Two methods can be utilized to break this habit. As with the two-ball pattern, practice close to a wall. This will not allow the ball to move away from you. The second method is to practice for a short time, with silk scarves. This will not only slow the pattern down, but also stop the natural reaction to avoiding the ball causing body damage. Unlike the juggling ball, the silk scarf will not cause body damage if landing on you or striking you in the head!

- Rushing the pattern resulting in two balls colliding. This is the mind panicking. Most pro athletes are very familiar with this one. It also can be found in manufacturing—excessive speed beyond line capacity resulting in excessive scrap and rework. This one is easy to stop. Go back to the two-and-half-ball pattern. Count the pattern to yourself: 1-2-space – 1-2-space – 1-2-space – 1-2-space. Now add in the number 3 in the space spot: 1-2-3-1-2-3-1-2-3. Switch to the three-ball pattern, but saying out load the numbers 1-2-3-1-2-3. This will have the effect of slowing the pattern down and creating a natural three-ball pattern.
- Coming to a stop before dropping. This usually occurs between the second and third toss. This habit is best broken with another juggler as a partner. Have a more experienced juggler stand next to you, and the two of you juggle three balls back and forth using only one hand each. This will trick the brain to continue the pattern, since the more experienced juggler will not stop until the drop. This technique will likely carry over to the three-ball cascade from the individual juggler.

Continue with the three-ball cascade until 10 consecutive tosses are reached. The remaining chapters in this section will utilize various statistical techniques to evaluate and improve the three- or two-ball cascade.

Figure 5.18 Advanced juggling—four-ball cascade.

Figure 5.19 Advanced juggling—three pins.

Optional advanced next steps for the advanced jugglers

For a challenge and after the following chapters are completed, consider attempting the following advanced techniques (Figure 5.18):

- The four-ball cascade (see Figure 5.19). All even number patterns past two-ball are normally not crossed from hand to hand. That is generally the case for the four-ball cascade. This is a pattern of two-in-one-hand for both the left side and right side. This is best to accomplish, like most processes, by breaking it down into smaller pieces. Practice a pattern of two-in-one-hand until 10 or more throws are accomplished. The second phase is to determine the pattern in the two individual hands. This can be columns on time in each hand, alternating one time or two upon one hand followed by two upon the other hand. Only attempt this after mastering the three-ball cascade.
- Three-ball cascade with juggling clubs. At the beginning of this chapter, links are provided to purchase juggling clubs. Three juggling clubs are not difficult but only should be attempted after the three-ball cascade is mastered. The degree of difficulty is amplified with clubs, but mostly as a result of the increased risk of injury, and

so it is best to start learning this step with gloves and protective head gear until 10 tosses are realized. This path of learning should follow the same steps outlined earlier for the three-ball cascade.

The forthcoming chapters will all outline the statistical and continuous improvement techniques to improve the art of juggling. Continue on to those chapters after developing a minimum of 10 tosses with the two- or three-object cascade with balls or silk juggling scarves. The latter are excellent at slowing down the pattern but are physically much more strenuous. Using either method or object combination is excellent for hand–eye coordination, brain food, or burning calories!

Bibliography

Beek, Peter J.; Lewbel, Arthur, The Science of Juggling, *Scientific American*, **273** (November 5), 92–97, 1995.

Draganski, Bogdan; Gaser, Christian; Buch, Volker, Neuroplasticity: Changes in Gray Matter Induced by Training, *Nature*, **427**, 311–312, 2004.

Polster, Burkard, *The Mathematics of Juggling*. 2003, Springer-Verlag, New York.

chapter six

The start of any process including learning to juggle

Mean, standard deviation, and the learning curve

Objectives

- Understand the creativity required to measure any process.
- Once measured, how to utilize base-level statistical concepts to start the modeling process.
- Determine what the data is telling you in terms of customer requirements.

The key to any process optimization: The right measurement

Let's take the juggling exercise learned in Chapter five. What is the metric that determines success? How it looks—good or bad? How it feels—terrible or terrific? Good day today, felt just right? Notice the trouble with these types of metrics: They are subjective and opinionated and not necessarily metrics to begin the continuous improvement journey.

Creativity in choosing measurement systems: Convert it to something measureable

What is the best way to measure success or failure in the juggling process? Let's try a system that is used frequently in equipment reliability systems: Mean time to failure. What is the equivalent in juggling? Number of successful tosses to drop (TTD). Very simply, count the number of tosses until the first drop regardless of the number of balls or objects utilized. This measurement system works for multiple industrial processes, administration processes, and even sports. Here are typical sport examples:

- Baseball pitcher. Number of strikes between hits. Hits is the failure; strike is a success. But what is defined as a hit? This is where creativity comes into play. If a line drive is hit directly to the center fielder, is it a hit even though it went for an out? If the ball is a slow grounder between third and the shortstop for a hit, is it a hit from a pitcher's perspective or a random bad luck hit normally an out? How the analysis changes depending on the argument!
- Baseball hitter. Number of line drives, balls, or deep fly balls between strikes. But perhaps not hits, as a broken bat hit by luck past the mound is likely not useful for statistical analysis but called a hit for official record keeping.
- Forward in soccer. Number of shots on goal, assists, or successful passes between any ball touches.
- Receiver in American football. Number of pass plays open to number of plays not open. This one may require creativity in determining what defines a success or "open." Is open when no defender is within 1.5m? What if the wide receiver is double-teamed? Is that a failure because the receiver is not open, or is that a success because it opens up another receiver on the team?
- Point guard in basketball. Number of trips down the court without a turnover up to the completion of the first pass. Number of successful passes without a turnover.
- Successful putts in golf. What is defined as a successful putt in golf? If it goes in? Hardly! Sinking a 9-m putt is a lot lower odds than sinking a 0.5-m putt. How about success is defined as any putt that ends up within 3% of the starting distance from the center of the cup. For example, a 9-m putt ends up within 0.45 or a 3-m putt within 0.09m.

Typical retail examples

- For a restaurant. Number of customers served within 10min of order before failure. Failure is defined as any customer order outside a 10-min window.
- For a grocery store checkout line. Number of customers through checkout before failure. Failure is defined as more than 2min from order ready to checkout but split between two subcategories—normal lines and express lines. Separating the two as separate processes with independent metrics would prevent what Dr. Donald Wheeler would classify in statistics as mixing apples and oranges together to create one metric. That one metric would likely misrepresent the process.

Banking example

- Number of successful transactions until failure. A successful transaction means the customer's entry into the bank and exit from the bank takes place within 3 min.

From nontypical processes

- Aircraft landings. What constitutes a failed landing? What defines a success? Is running off a runway a failure? Probably not the correct measurement as it may take years before a failure, and it may be more luck than any other reason. But in terms of the passengers, what constitutes a failure? What about the impact at landing as the measurement of success? Can it be measured? It can be, with some creativity. If it can be measured, is there a threshold at which that customer rating of the landing changes from rough to acceptable on a typical ranking scale? The measurement we look for to determine success of a pilot is then the number of acceptably soft landings between rough landings.
- For an industrial machining process. Number of cycles to machine breakdown. Maybe, but that may not cover the number of successes to quality breakdown. So, change the measurement to be more representative of the success of the process—number of parts made within ±3 standard deviations of the target to failure and on time, in between failed parts not meeting those criteria. This will add a dimension to producing a quality part and equipment breakdown.

Mean and standard deviation for tosses to drop as a training technique for process optimization

Start with a two-ball cascade for all students. Have students begin juggling, recording number of TTD. Do this for anywhere between 7 and 15 complete cycles. What is the mean? What is the standard deviation? What does the standard deviation tell each student about the process? What if your goal was to consistently hit a minimum of 10 cycles? What is the capability of the process? Is the process in control (Chapter seven)? What to do if it is not in control?

These are questions typical of a process optimization in a manufacturing process or transitional operation. All questions are best answered not just by engineers and technicians, but also by operators, clerks, nurses, and many other frontline personnel.

Start with the basics. What is the mean and standard deviation of this process? Typical mean and standard deviation calculations are as follows:

$$\text{Standard deviation} = \sigma = \sqrt{\sum_{n=1}^{n} \frac{(\bar{X} - X)}{N}}$$

$$\bar{X} = \frac{X}{N} = Mean$$

This calculation is simple to do on a standard calculator or spreadsheet program.

Typical tosses to failure of a juggling process during the learning phase was covered in Chapter five.

Is this juggling performance a success? To answer that specific question, we need the most important information during process development: Who is the customer, and what is the specification? In this situation, the customer is likely to be the student themselves. The student can develop a minimum target necessary for their own satisfaction, or they can develop what they would consider an expectation of minimum competency. Determining the customer expectations for other processes is often difficult, because mostly it is not determined until after the product is in operation. In this case, let's utilize five TTD as the critical minimum specification. For this example, there will not be a maximum specification as we would be perfectly happy if the juggling went on to 1,000 or 10,000 tosses (Note: it is unclear who holds the record for the number of TTD for three ball, but the record for juggling five pins is 45 min, set by retired professional juggler Anthony Gatto. The number of tosses is estimated at 3800 before that infamous drop).

Process capability calculation: Number of standard deviations

There are two typical calculations used for determining process capability. The first is a straightforward number of standard deviations from the minimum or maximum specification. The other is a capability index, commonly shown as CPk, CPl, or CPu. For this exercise, the number of standard deviations from the maximum specification and the capability index of CPu is not defined, as there is no upper specification.

For the number of standard deviations from the lower specification based on the original distribution used in Figure 6.1:

$$\frac{(\text{Mean} - \text{Lower specification})}{\text{Standard deviation}} = \frac{(10-5)}{3} = 1.66$$

So our process capability measurement is 1.66; we are 1.66 standard deviations above the lower specification (Figure 6.2).

The process is assumed to follow a normal distribution for this analysis.

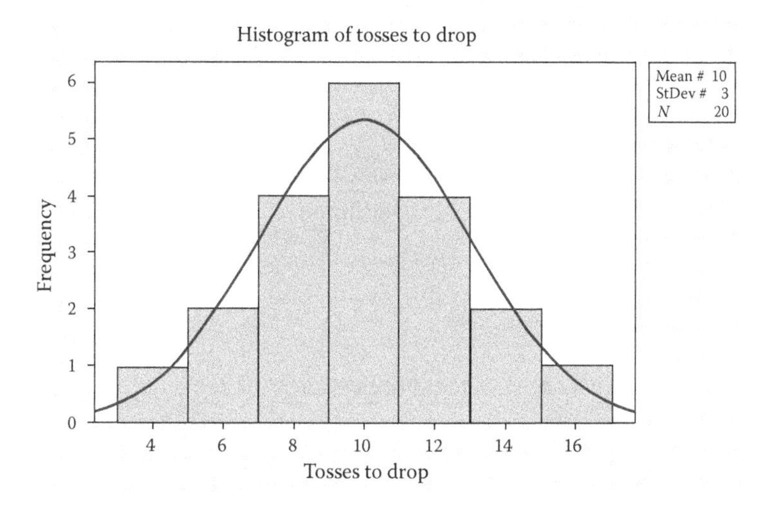

Figure 6.1 Tosses to failure for a mean of 10 and standard deviation of 3.

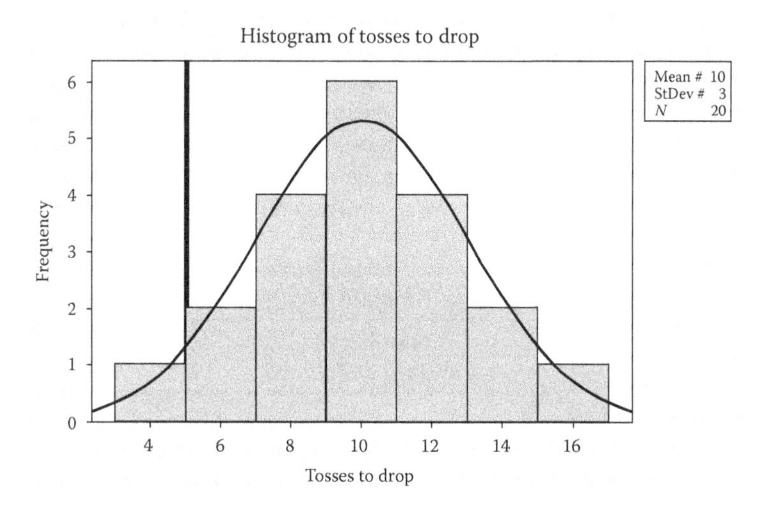

Figure 6.2 Process with minimum customer specification.

Process capability by capability analysis

A second and somewhat more typical analysis of process capability is by capability index or CPk, CPl, and CPu, which are just a few of the many calculations. The typical calculation is as follows for CPl:

$$\text{CPl} = \frac{(\text{Mean} - \text{Lower specification})}{3 \times \text{Sigma}}$$

For the upper specification:

$$\text{CPu} = \frac{(\text{Upper specification} - \text{Mean})}{3 \times \text{Sigma}}$$

CPk = the lower of CPl or CPu
For this initial juggling exercise:

$$\text{CPl} = \frac{(\text{Mean} - \text{Lower specification})}{3 \times \text{Sigma}} = \frac{(10 - 5)}{3 \times 3} = 0.55$$

Typical guidelines for CPk and number of standard deviations calculations are provided in Table 6.1.

So in our calculation of the juggling exercise, the process capability (CPl) to the lower specification of 5 is less than desirable. The goal now is to improve the capability of the process by continuing the Six Sigma process. We've been following the process already without discussing explicitly—the generally accepted process includes five steps. The steps are as follows: (1) Define, (2) Measure, (3) Analyze, (4) Improve, and (5) Control. We have already defined our metrics/measurements as the number of TTD and then measured our current process. Steps 3 and 4 are the Analyze and Improve steps of process improvement. There are several phases within Analysis and Improve. Here, three

Table 6.1 Acceptable capability guidelines for many processes

	Number of standard deviations	CPk
Fair	3	1.0
Good	4	1.33
Excellent	5	1.67

concepts will be covered in detail for this process: analysis of control or SPC, understanding of the equipment reliability function, and design of experiments.

Additional exercises

As additional starting exercises, the following has been found to be a useful set of questions to start the continuous improvement process:

- What is the new process capability if our minimum specification is 8 tosses to failure instead of 5? Should this be 10 tosses to failure or 12 tosses to failure?
- Can CPk ever turn negative, and if so, under what circumstances?
- The earlier calculation was for tosses to failure. What changes for cycles to failure? Determine an equivalent process that is judged in cycles to failure and not parts to failure.
- Which hand is dropping more? Start the process over, and record the drop by hand. Is there a significant difference? (Note: The concept of significant will be explored and shown by experimentation in Chapter eleven.)
- If the answer to the earlier questions is yes, there is a difference, but where is the opportunity for improvement? In the throw or catch? Develop a secondary experiment to answer that question.

What happens if the height of the throw is changed? Does that factor affect the tosses to failure? Design and initiate a controlled experiment for the height of the throw. What is the measurement system for the height of the throw, and how accurate is the system to the expected variation of the process? This will be a great introduction to the Chapter nine—the concept of design of experiments.

Author's note

This section is a very basic section for understanding the continuous improvement process. But it moves it away from being a strictly academic memorization of formulas to something that students can appreciate, feel, and apply. Properly applied, this can also start the conversation on other topics such as confidence intervals, positional control, etc. The next sections will demonstrate what to do when the customer (the student) is not satisfied with the outcome. The experimental process will be introduced and expanded in a series of designed experiments and through analysis of positional control.

This exercise has a side benefit, as the learning steps for juggling are the same for almost any process. It can follow the Plan-Do-Check-Act cycle or Six Sigma cycle. Most students' confidence will increase in other areas after learning this new sport or process.

Bibliography

Benbow, Donald W.; Kubiak, Thomas M., *The Certified Six Sigma Black Belt Handbook.* 2005, American Society for Quality Press, Milwaukee, WI.

Breyfogle, Forrest W., *Implementing Six Sigma*, 268–275. 2003, John Wiley & Sons, Hoboken, NJ.

Munro, Roderick A.; Maio, Mathew J.; Nawaz, Mohamed B.; Ramu, Govindarajan; Zrymiak, Daniel J., *The Certified Six Sigma Green Belt Handbook.* 2008, American Society for Quality Press, Milwaukee, WI.

Snee, Ronald D.; Hoerl, Roger W., *Leading Six Sigma: A Step-by-Step Guide Based on Experience with GE and Other Six Sigma Companies.* 2003, Pearson Education Inc., Upper Saddle River, NJ.

chapter seven

The basics of using the control chart to improve the juggling process

Objectives

- Understand the basics of statistical process control (SPC).
- Understand how to apply a simple SPC control chart to the process of juggling.
- Describe three common out-of-control signals and what they may mean.
- Understand how adjusting an in-control process is likely to increase variation.

The basics of the Shewhart statistical process control chart

SPC was first developed by Walter Shewhart in a 1918 one-page paper written while he was working at Bell Laboratories. That one-page paper had in it what is now believed to be the very first control chart. That one-page paper, years later, likely set the stage for the revolution in changing most processes from inspection based to improvement based. A good portion of Six Sigma is based on Shewhart's theory and expanded by other notable experts in this area such as Dr. Edwards Deming (who coauthored several papers with Shewhart) and Dr. Donald Wheeler. Dr. Shewhart went on to write the now very famous book on this topic, *Economic Control of Quality of Manufactured Product*. Published in 1931, this groundbreaking book likely addressed, for the first time, the theory of overadjustment or how haphazardly adding specifications to a controlled process for adjustment purposes ultimately degrades most processes. That concept will be shown at the end of this chapter.

So from a high level, what is SPC trying to accomplish? There are multiple methods of defining it; however, basically what Dr. Shewhart says is that if a process is within approximately ±3 standard deviations of the mean with no unusual patterns (more on that later), the process is in statistical control. If the process needs improvement, another method will be required to improve the process. These improvement methods will be shown later, but for now, address only those data points that show an unusual pattern—leave everything else alone. But what if the process is not meeting requirements and has no unusual points? Same statement—leave it alone until more advanced sections in later chapters explain a better methodology. But why is the use of SPC documented over the years as being an effective method of improving any process? Because most processes, especially those that have never had a control chart on the process before, are not in control. There are multiple out-of-control points that, if utilized as feedback signals, are significant and sometimes huge opportunities for long-term improvement. One of Dr. Deming's quotes probably best summarizes this area:

> a state of statistical Control is not a natural state… It is an achievement, arrived at by eliminating one by one, by determined effort, the special cause of excessive variation.
>
> **Dr. Edwards Deming**
> *Out of the Crisis*

A process is rarely in control without work done to keep it under control. Investigating out-of-control signals and permanently correcting the conditions causing those signals is not a natural activity. It is also not a natural activity to leave a process alone if it does not give a clear, beyond reasonable doubt, signal that it is not performing. Messing with such a process is commonly referred to as overadjustment. SPC, if used properly, and as has been shown by Dr. Shewhart and thousands of companies that have put the effort into SPC, is effective and as will be demonstrated, also works for a simple process like juggling.

The basics of statistical process control

SPC is the time-oriented assessment of process performance. It answers the questions: Is the process predictable over a unit of time? Is there a significant, and beyond reasonable doubt, change in the process given the recent history? By itself, it does little if anything to answer the question as to why the process is not in control. It only points out to the responsible interested party; there is likely an opportunity for improvement. Does it signal a point for adjustment? With a very few exceptions, most times no. Although rare, most out-of-control conditions that may require adjustment are associated with environmental factors or tool wear (see Chapter eight).

The generation of the control chart typically would start with a Gauge R and R study. A Gauge R and R study is used for three primary purposes: to determine if the measurement system is acceptable for the process in question, to assess the stability of the gauge, and to determine an acceptable method of holding the gauge accuracy. For this process, the measurement system for the physical counts to drop will be considered acceptable to proceed. This can be verified by a typical Gauge R&R study. For this exercise, the measurement of system for this process, that is physical counts to drop, will be considered acceptable to proceed.

There are multiple variations of the SPC chart. The variation in type depends on two primary factors: the underlying distribution and what Shewhart called natural subgrouping. To explore further:

- What is the underlying distribution? Is it the normal distribution? Is the data skewed by a natural process such as a chemical phase change? Is there a lower bound in the data typical of data that cannot go negative in value? The underlying distribution typically determines the type of chart. But for our example, the distribution will be the normal or Gaussian distribution.
- Rational subgrouping depends on three factors: the context of the data, how the data will be used, and the question to be addressed by the chart. If the chart is designed to represent a small proportion of the population, the rational subgrouping will be significantly different from the samples representing a significant proportion of the population. Refer to Dr. Donald Wheeler's book, *Advanced Topics in Statistical Process Control*, for more information on this topic. For this exercise, a rational subgroup of 1 will be used, as typically used in the individual and moving range control chart.

The typical three primary conditions for out-of-control

For this exercise, only three conditions will be used to identify an out-of-control process. There are many (by last count, most statistical software packages had a minimum of eight test options). These three will normally not only signal a change, but also provide some evidence of where to look for the cause of the condition. This will be outlined with the juggling example and followed up by what typically happens in industry or transactional area. Three conditions signal a significant, beyond reasonable doubt, change in the process. They are as follows:

- A single point above or below the upper or lower control limits. This typically is the easiest to understand. Above or below the ±3 standard deviations from the mean is a rare event and signals an opportunity for improvement. A point falling outside of the limits is an opportunity for analysis of the data point. Analysis of the data point will lead to improvement if, after understanding the underlying cause, long-term corrective measures are taken.

- Eight in a row above or below the mean. Why this condition? This normally identifies a complete shift in the process, pointing to a change in input parameters such as raw materials, suppliers, etc. This exercise will show how an induced condition can be used to change the process.
- Six ascending or six descending. Why this condition? As will be shown in a later chapter, this condition normally identifies a condition of wear out or learning curve. Both these conditions, commonly found in manufacturing and transactional processes, are typically covered in equipment reliability examples but will be demonstrated in this exercise.

The juggling demonstration of statistical process control

Once again, the metric used to measure the process, that is, number of successful tosses between drops, will be deployed. It is easy to count and is an excellent demonstration for an elementary control chart. For this exercise, even with a lower bound of zero, we will utilize the normal distribution and the individual and moving range chart (I-MR chart). Why the I-MR chart and not the $\bar{X}$-R chart or an attribute-based chart? We will be measuring every attempt or cycle to failure and the out-of-control conditions identified. This follows very closely Dr. Donald Wheeler's statement, "the question to be answered by the chart." The question to be answered by the chart will be "analyzing as much of the data as possible in time order, what information can be obtained from the out-of-control conditions to improve the process?"

The first view of the process, an isolated out-of-control point, and what information can be obtained

After learning the process in Chapter five, two-ball pattern for tosses to failure in this example, with an example student, looks like that shown in Figure 7.1.

Basic terminology for the control chart are as follows:

- The upper and lower control limit is theoretically based on ±3 standard deviation from the mean ($\bar{X}$). In this exercise, they are represented by the upper control limit (UCL) of 17.5 and lower control limit.
- Although not shown in this exercise, the range chart represents the variation of the process. A change in variation for the range will

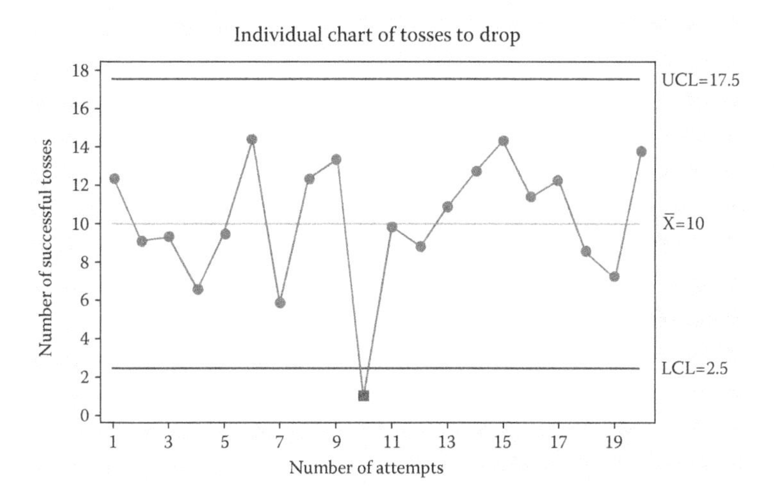

Figure 7.1 Tosses to drop for two-ball cascade.

commonly appear as out of control on the range chart. For more information on the range chart, refer to Dr. Donald Wheeler's book, *Topics in Statistical Process Control.*

Analysis of the control chart shows one out-of-control condition. That point was at toss number 10 and resulted in dropping the second ball (one successful toss). What caused the drop? What unusual condition was present? How do we use that point to improve the process? In this situation, the arm was caught on the pants pocket as the left hand attempted to catch the toss from the right hand. Easy to fix? Yes it was—change the clothing to athletic wear, thus eliminating the cause. Found otherwise? Maybe, maybe not. But notice how the control chart clearly identified the outlier. Simple and to the point. Stop, and identify the cause by investigation; permanently eliminate the cause of the condition, thereby permanently improving the process. Simple and easy to apply, but not likely an opportunity if the condition was ignored or the condition was not permanently eliminated by adjusting all future throws or, as so often occurs, adding a sign on the process to watch out for the pants pocket.

Is this typical of a manufacturing or transactional process? Likely, as most processes are subject to isolated conditions and spikes. Some examples of this are given here. A manufacturing process with the top board of the bulk process warped because it was sitting outside exposed to the elements for 2 days. A slip in the coupling after it was hit during a specific run caused by the top board in the bulk process being warped.

A spike in the time to process at a bank caused by a teller on break and the dismissal of a group of people at a convention that was meeting in the building. A spike in the attendance at the local theater company's Sunday performance after a special 1-day online sale. A spike in the power supply at 2:00 pm in a Mexicali Mexico manufacturing operation when the power grid changes induced by the rate change at 2:00 pm.

How best to simulate this in a classroom setting? As the students are juggling, take a loud mechanism, and turn it on unexpectedly. The jarring noise will undoubtedly result in ball drops. Time it so that several are at the start of the run. This will likely show up as an out-of-control condition for one spike only.

Special cause of variation or nontypical behavior—Eight in a row above or below the mean

This will generally represent a shift in the entire process associated with a change in supplier or supplier material. Why 8 in a row and not 3 or 4 or 12? Following Chapter two on the multiplication principle, what is the probability of getting a single point above or below the process mean? Hopefully, students will understand it to be 50%. What is the probability of eight heads in a row all independent? $.5^8$ or .0039, which is very close to the same probability of a point outside the upper or lower control limit.

Developing the control chart based on the baseline analysis for the next 20 points results in what is shown in Figure 7.2.

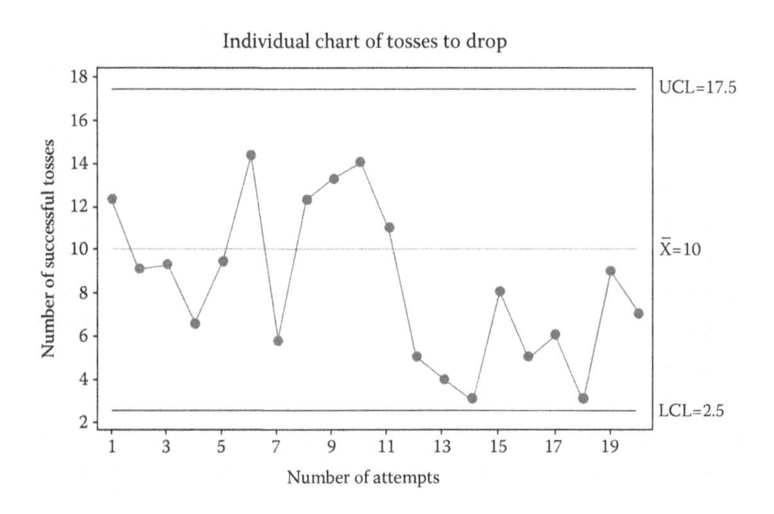

Figure 7.2 Tosses to failure—eight in a row below the mean.

This is a clear signal of a process shift in the mean, with the last two points clearly signaling an out-of-control condition. Investigation of the process revealed it to be a change in environmental conditions—the room heater fan turned on. For example purposes, this process shift can be set up by multiple methods such as the following:

- Turn the set temperature in the room down or up when the first 10 are completed. This will turn the fan on in the room, changing the airflow.
- Purchase an external fan, and turn it on at the appropriate point close to the 10th toss.
- Have all students change juggling balls to a lighter ball after the first 10 tosses. This inadvertently results in a decrease in efficiency commonly associated with having to learn a new ball shape and weight. This is an excellent example of an undocumented change in incoming product supply.
- Do the first 10 before lunch and the second 10 after lunch. This will likely result in a drop in efficiency after lunch. This is similar to a change in manufacturing lubrication such as specific gravity or viscosity or a change in the dimensional setup brought about by a change in running temperature.

How does this relate to a real-life process? In numerous ways, but here are a few that have been found to be prevalent forms of total process shifts:

- Two raw materials are utilized, with one of lesser quality. Unknown to the operations, the process shifts from one supplier to the other.
- Two operators run the line but with different skill levels. One reports off for the day.
- Preventative maintenance is performed at the break, but the press bearing is changed out. The new bearing has a different drag on it as the material passes through the side supports, resulting in the material skipping as it goes through the heater tunnel. This results in an increase in exit temperature, which changes the line speed program requirements as it was designed on a certain heat input.
- A new clerk is added on to the register at a restaurant with limited experience. Training time was not sufficient. Time to process invoices suddenly increases.
- A runway is closed at the local airport for maintenance. Only two runways remain open instead of the normal three. This results in a larger-than-normal wait time for takeoff and a spike in airlines' deviation from scheduled departure.

Nontypical process behavior—Six in a row trending up or down: The wear out

The final critical, nontypical, or special cause is the trend with six in a row up or down. What is this signaling? It is normally a wear-out condition. This can be a machine wear out or a human wear out. This condition can best be demonstrated in the juggling exercise after some proficiency has been established with the two-ball or three-ball cascades. This can be done with an existing process at the end of the shift or accelerated by switching to weighted balls. A typical control chart establishing this pattern is shown in Figure 7.3.

How is this factor found in real situations? Here are several conditions that might be classified as wear-out:

- A machine bearing reaches the end of the cycle resulting in an increase in vibration, which in turn results in an increase in part diameter.
- Physical unloading rates drop at 3:00 pm for a dayshift operation.
- Ability to take in new material during a lecture plummets after 1 h without a break.
- A spike in the accident rate at 4:00 am in many industrial operations owing to operator fatigue.
- A spike in the accident rate late in the day on the first business day after the clocks are turned back 1 h in the spring—a common event in many countries.

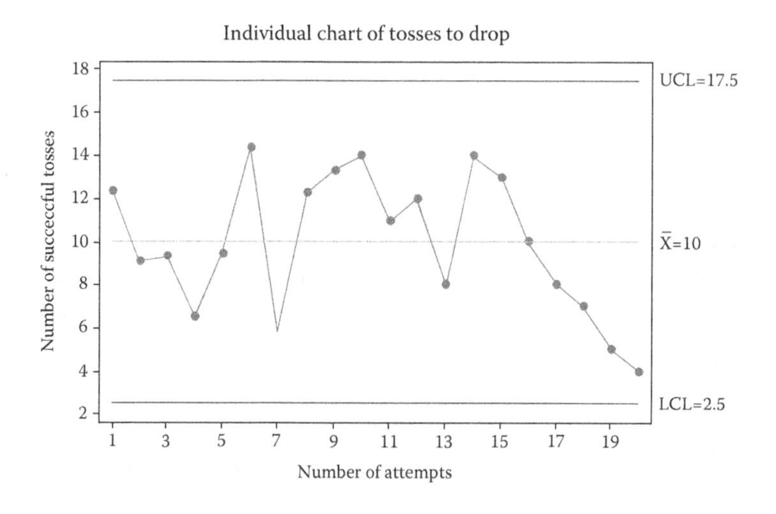

Figure 7.3 Control chart of special cause—wear-out.

- A decrease in the checkout rate at a typical retail store after employees are on the line for more than 8 h.
- The computer error entry rate at an online processing center during the seventh and eighth hour of a work shift.

The wear-out factor will be explored more in the next section on the bathtub curve.

How to not use a control chart—Adjustment and the negative consequence

Dr. Shewhart addressed this in his first paper and his book on the topic; Dr. Deming best learned how to demonstrate it. Shown below is the negative effect of adjusting an in control process or, as Dr. Deming used to call it, mixing up common cause variation with special cause variation. Two graphs explain the Deming funnel experiment (Figure 7.4).

The first graph (Figure 7.5) is from Dr. Deming's funnel experiment. In this exercise, Deming takes a funnel and drops marbles through it to a matrix with a target centered over the end of the tunnel. There is minimum variation in the first graph with the target not moved. The second (Figure 7.6) is the extreme case of moving the target after every drop in the opposite direction from where it landed—trying to keep it on target by adjusting every drop. Notice what happens to the process and the variation—it is worse and not by a small amount. This effect is a common industrial occurrence most notably brought about by management pressure applied to the operators or operators themselves in a fruitless attempt to improve the process.

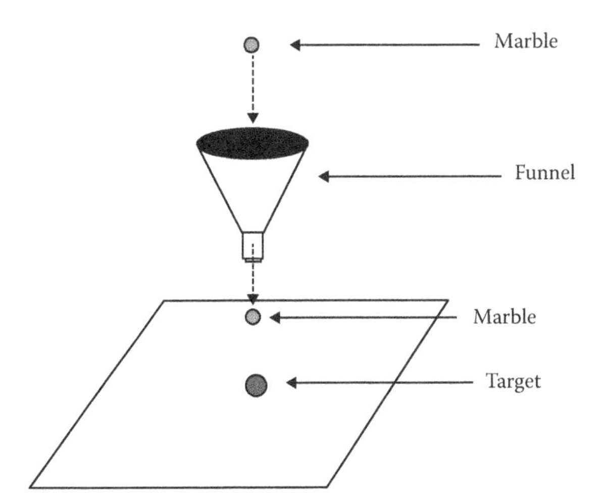

Figure 7.4 Deming's funnel experiment.

Funnel drop position remains the same

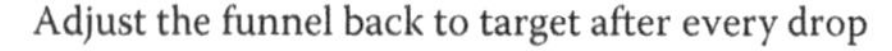

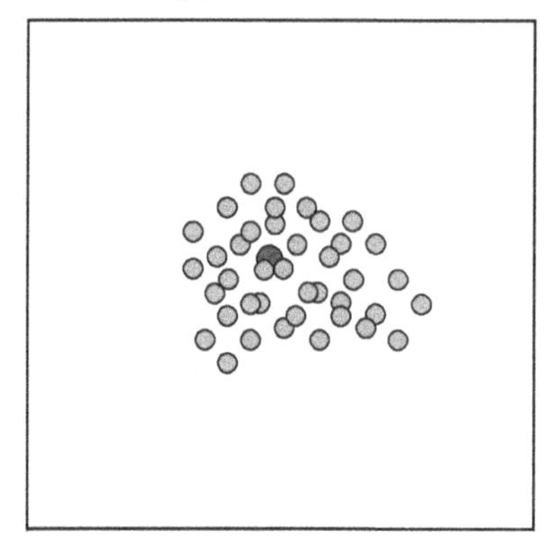

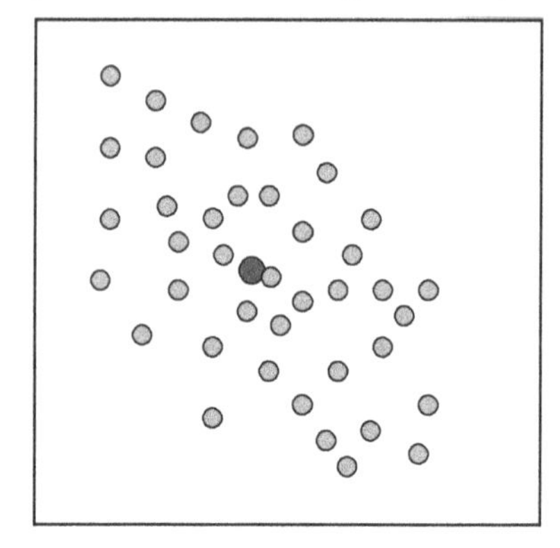

Figures 7.5 and 7.6 Increase in variation by adjusting a process.

So if there is an attempt to place customer or management specifications or desires on the control chart, it will not only result in any positive effect but will also most likely make the process worse. Dr. Wheeler had the best explanation of this in his classic book, *Understanding Variation: A Manager's Key to Understanding Chaos*. Under pressure to perform (specifications on the control chart), people will distort the process or change the data. Typical control charts with specifications or goals will commonly look like the one shown in Figure 7.7. This is an example pattern of a process likely in control but not meeting customer specifications. Notice the "cliff" in the process. For this application, the supervisor requested an explanation note from the technicians for any data point above 30. What is the reaction to forced compliance? Change the data, or distort the data. In this case, a second sample was taken and sent to the lab for analysis. If the second was under the specification of 30, it was accepted. If the second was not, the data was changed. The process was recorded as meeting customer specification but was clearly altered as a result of the specification—a clear case of not listening to the voice of the process as it was distorted by the specification.

Any attempt to adjust a controlled process is likely to result in degrading the performance. But this one is hard for most managers to understand. But what if the process is in control but is not meeting customer specifications? See the previous analysis; adjusting the process is likely worse than a waste of time as it will only make it worse. In the conditions of in control but not meeting customer expectations, further understanding of the causes of variation is required. That will be explained in further chapters.

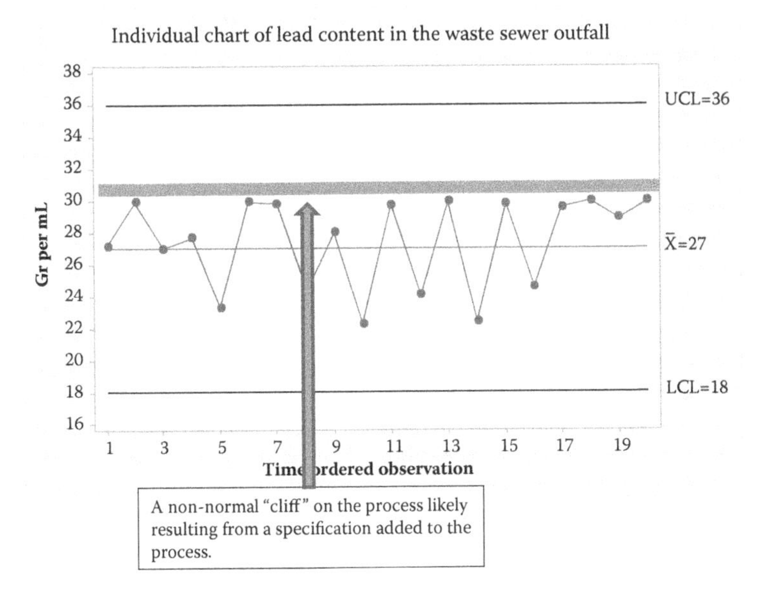

Figure 7.7 Control chart of a tampered process.

Author's notes

- Control charts work great in many applications if the intent is understood—identification of nonnormal variation in a process and the permanent elimination of the cause of the nonnormal variation. This will ultimately reduce the variation in the process, leading to a much more robust process. Shortcuts in those steps, fabrication of the data, or haphazardly adjusting the process will only degrade the process. It is better not to start corrections than attempting to correct the process without full understanding and commitment.
- Control charts have also been called process behavior charts primarily by Dr. Donald Wheeler. I tend to agree with Dr. Wheeler, as this definition more accurately describes the chart as defining the process behavior rather than a control mechanism. But for the rest of this book, they will be referred to using the more conventional terminology.
- Although around for years, SPC has its limits. A poorly designed process remains a poorly designed process for which SPC will not likely be the magic cure. For a poorly designed process, a poorly implemented SPC will result in the process staying the same at best or will result in the process becoming significantly worse (worst-case scenario). For a poorly designed process, there are

much better systems to utilize than SPC alone. See the following chapters for more details on improving the process through experimentation.

Bibliography

Breyfogle, Forrest W., *Implementing Six Sigma: Smarter Solutions Using Statistical Methods*. 2003, John Wiley & Sons, Hoboken, NJ.

Deming, W. Edwards, *Out of the Crisis*. 1986, Massachusetts Institute of Technology, Center for Advanced Engineering Study, Cambridge, MA.

Shewhart, Walter A., *Economic Control of Quality of Manufactured Product*. 1980, American Society for Quality Control, Milwaukee, WI.

Wheeler, Donald J., *Advanced Topics in Statistical Process Control*. 1995, SPC Press, Knoxville, TN.

Wheeler, Donald J., *Understanding Variation: The Key to Managing Chaos*, 2nd Edition. 2000, SPC Press, Knoxville, TN.

chapter eight

The reliability function or bathtub curve as demonstrated by juggling

Objectives

- Begin to understand the reliability function or bathtub curve.
- Understand how the sport of juggling can be used to teach the bathtub curve.
- Where and how the bathtub curve shows up in real operations.

Demystifying the bathtub curve for equipment and personnel

Let's look at the typical graph for an equipment lifecycle (Figure 8.1). What does it tell us about most human or machine processes?

What is it really saying? There are three sections that will be reviewed, but the first and last are critical to equipment, personnel performance, and reliability.

There is a typical break-in period for most processes. That is the first section on the graph appropriately named and labeled as Infant Mortality. In this section, equipment will historically have a lower reliability than Section two. There are entire textbooks on Section one. Actual examples will be described at the end of this chapter, but for now, consider this the typical break-in period or warm-up time. This process is experienced in almost all sports, academic areas (American Society for Quality's exam development process will typically include easier test questions at the start of the exam in recognition of this area of the curve), growth functions (although decreasing, the human body mortality rate follows the infant mortality curve, especially for ages 0–4)—hence the name, and many natural functions (think of this the next time you rake leaves in the fall addressing the remains of the wear-out cycle and how fragile those leaves were 8 months ago). Notice the critical characteristic of this section—the decreasing failure rate with time. This section can be the

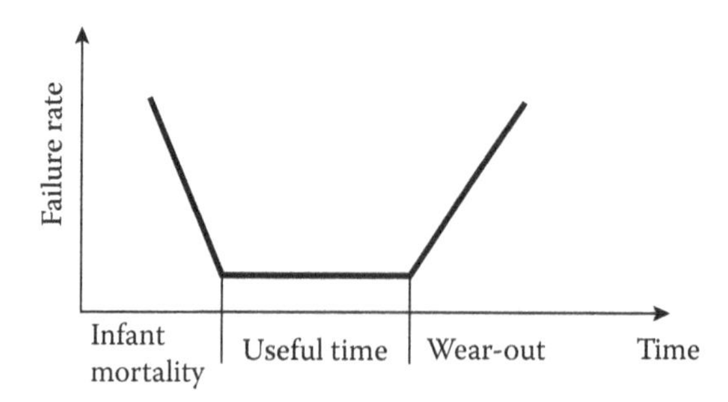

Figure 8.1 Equipment reliability function graph.

most frustrating as a manager or operator, and many times this is the easiest of the three to reduce or eliminate through a technological or procedural change.

The second section is the useful time or constant breakdown frequency in equipment reliability or human performance. Characterized by the lowest frequency of failure, this section can be modeled by various functions such as the exponential or various Weibull distributions. The prior section SPC charts were developed based on the useful section or constant failure of the equipment reliability function.

The last section is the wear-out section. In terms of human growth functions, this is old age. In equipment functions, this is characterized by the beginning of the end of useful life. This is the automobile at 500,000 km, the human knee with <1 mm of cartilage left, the instructor at the end of a 2-h lecture, the frequency of accidents during the last hour of a shift, etc.

How to explain by juggling

For recall 1 month to 10 years later, refer to Chapter one's guideline: relevant, useful, and interesting/shocking. Is the concept of the bathtub curve relevant? If explained in the context of real-world examples—probably a seven. Useful? Depends on the audience or students—if it is used for exams, work-related examples or something in their own lives—probably a seven. Interesting/shocking? Unless an exercise is developed for this concept, interesting/shocking is probably a two.

This is an example of an in-class juggling exercise that should showcase the equipment reliability function or the bathtub curve over a complete cycle.

Start with the two-ball cascade at the start of the day or classroom time. Utilize the established mean and standard deviation from the prior

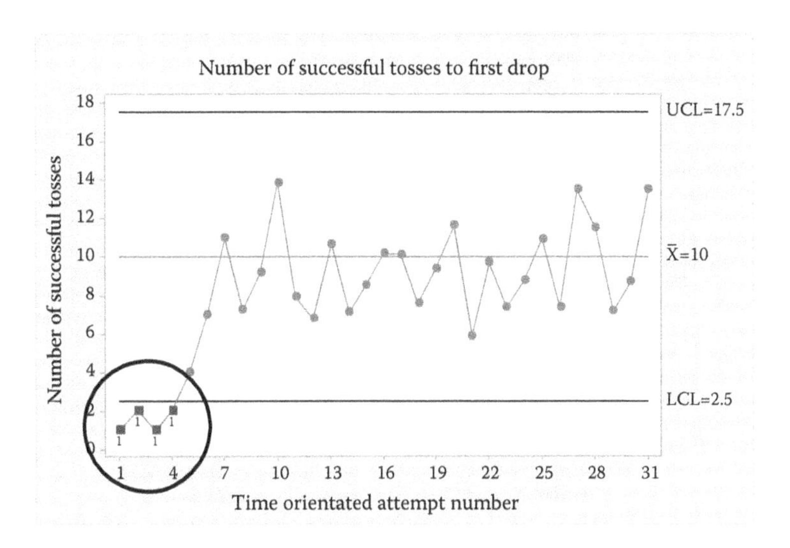

Figure 8.2 The control chart showing the warm-up or break-in section.

exercise. The typical control may look as shown in Figure 8.2, and the infant mortality section is in the circle.

The data gathered during the warm-up time is likely to be less than the mean from the last work session, but increasing in performance or time to drops. This period is historically analyzed differently from the constant or useful section of the graph, but can lead to long-term improvement opportunities. In manufacturing, innovations learned in this section can lead to developments that will commonly have a positive effect on the other two sections. Examples of this in manufacturing might be a different lubrication developed in a casting process for the period when the die is brought up to temperature and that new lubrication having a positive effect on the entire cycle of operation. A new mistake-proofing technique developed in a clerical process for the first hour of the day when data shows a high but decreasing frequency when implemented has a positive effect on the entire operations cycle. An improved alignment technique for a coupling that was prone to failure the first hour after installation had a positive effect on zone three, the wear-out zone, as the cycles to failure increased.

In the classroom juggling example, the drop in tosses to drop in the first hour can be brainstormed for potential cause. Opportunities for improvement found during the warm-up phase are usually amplified and thus easier to identify. Figure 8.3 is typical of what would be found in the juggling example or any other similar process; the opportunity for improvement is amplified in this section.

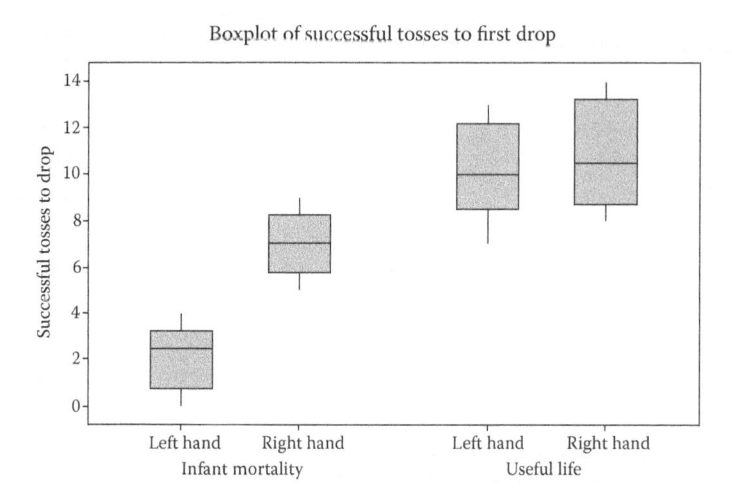

Figure 8.3 Graph of left-to-right drops during warm-up and useful life.

The useful life to wear-out transition zone

The transition zone from useful life to wear-out is the second critical zone. This is the zone that historically can be predicted and usually follows the normal distribution. This section is the critical section in predictive maintenance, because it can significantly convert unscheduled downtime to more efficient, scheduled downtime. In nonmanufacturing operations, the transition zone is commonly associated with fatigue, either physical or mental. Once recognized, this area is commonly associated with techniques to reduce or eliminate failure: increased frequency of breaks, changes in the typical 8-to-5 scheduling, and other areas such as diet and exercise changes to reduce fatigue.

Juggling can be utilized as a technique to better understand and improve the performance associated with wear-out. To start, with no other changes identified, induce the wear-out point by having all students juggle for 10–20 min nonstop. Record the cycles to drop and add to the control chart. The typical control chart will be as shown in Figure 8.4.

A demonstration on how to extend the wear-out cycle

There are multiple ways to extend the wear-out cycle. In manufacturing processes, a designed experiment (Chapters nine and twelve) will commonly locate an opportunity for major improvement. After an

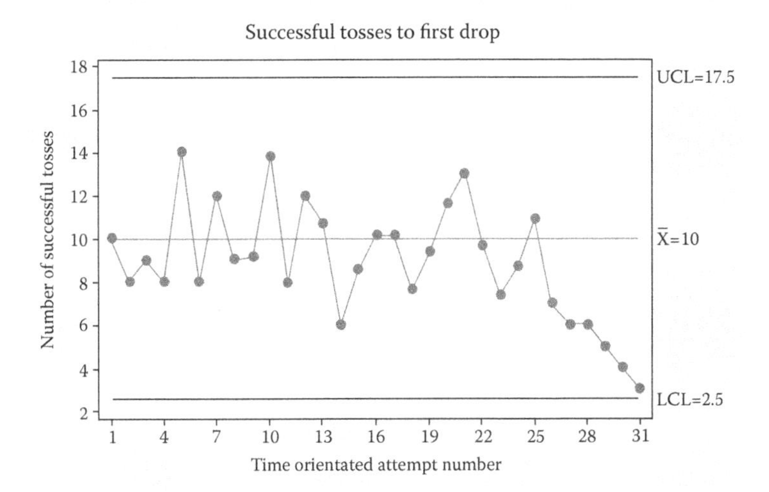

Figure 8.4 Control chart of wear-out condition.

improvement in the cycle found during this method, the next step usually involves a change in design or product.

For the juggling exercise, consider the following as options to best extend the cycle:

- Add lighting to the room. This is a great analogy to an actual non-manufacturing environment.
- Add a break during the useful life or constant breakdown section. This is in effect restarting the cycle and not changing the life cycle, but it will have a similar outcome.
- Fuel or food. Add fluid or carb intake before the downward cycle begins.
- Slow the cycle down by adding height. This is probably the least effective but has shown on occasions to extend the wear-out point as the frequency is slowed. This is analogous of purposely slowing the manufacturing process down in recognition of this sometimes unavoidable consequence.

The two graphs should look similar to that shown in Figures 8.5 and 8.6. Notice that the wear-out cycle is delayed but not eliminated in the second graph. The inflection point is rarely eliminated, but it can be delayed significantly.

The equipment reliability function or bathtub curve is a natural function found in most of all processes—manufacturing, nonmanufacturing, and even in nature. Properly recognized, it can be one of the most effective ways to identify opportunities for improvement. Not recognized, it can

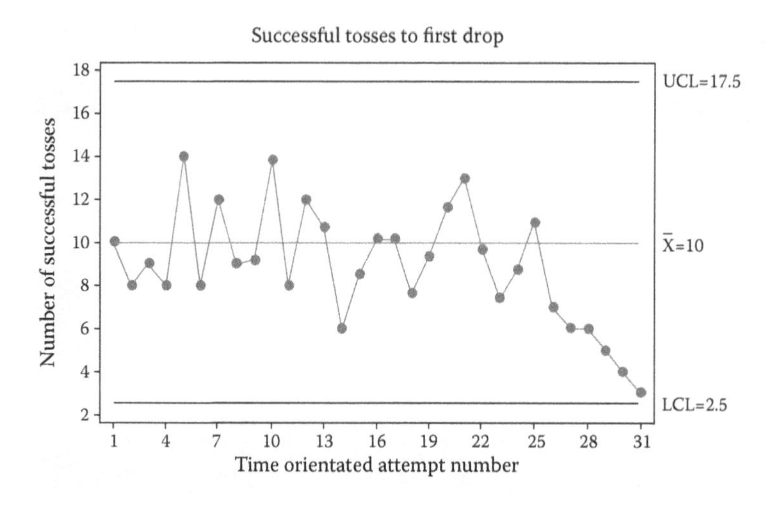

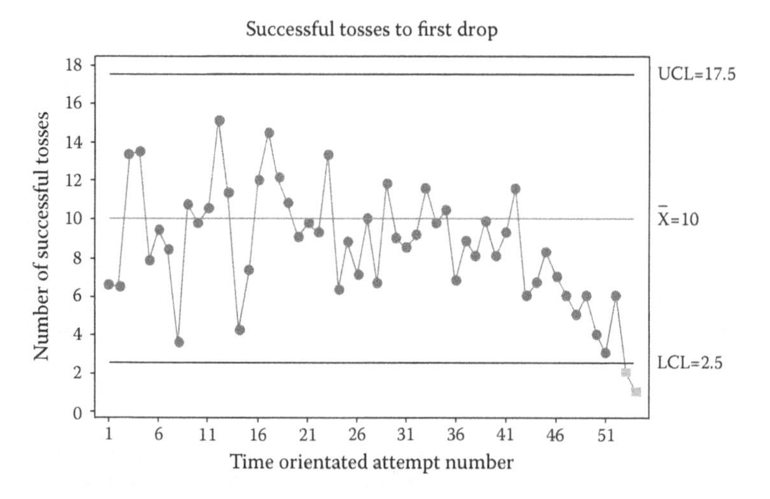

Figures 8.5 and 8.6 Graph of two processes: normal conditions and attempts to extend time to wear-out.

lead to an endless cycle of frustration. How best to improve on the cycle is one of the steps for continuous improvement. To recap the three phases:

- Infant mortality, warm-up, or break-in is the first phase. It is commonly associated with a higher frequency of failure or breakdown than the second section, but the rate is decreasing or getting better. Properly recognized, this area can be used to improve the entire process in addition to this area.

- Constant failure is the second phase. This is the area most associated with processes. This normally is the lowest and most consistent failure rate. For the juggling exercise, this will be characterized by consistent variation with relatively low mean in comparison to the other two areas.
- Wear-out is the final phase. For manufacturing, this is normally characterized by an unplanned breakdown. For nonmanufacturing, this is characterized by an increase in error rate, slowdown of efficiency, and overall drop-off in effectiveness. For the juggling exercise, this can be seen as a steady decrease in the tosses to failure. This can also be broken by changing the process as characterized by environmental or physical change.

Further classroom exercises

- Test out other areas of infant mortality. What other areas in the home, dorm, or classroom follow the bathtub curve or wear-out cycle? (Example: All automobiles have many areas of the infant mortality. This can be break pad/rotor wear-in until point contact is reduced to surface-to-surface contact. Clutch contact points. Years ago, all cars required a "1000 miles check-up," which was used to address the infant mortality curve.)
- Analyze other sports such as baseball, soccer, or basketball. Do they follow the bathtub curve? At what cycle? Hourly during the game? (Hint: Look at the pitcher's strikeouts by inning.) During the season? During a career? Look at the home run rate of Mickey Mantle and Roger Maris by year across their career and see the bathtub curve. Also, notice Sammy Sosa's home run totals by season. Did they follow the bathtub curve? If not, why?
- Develop the drop by left-hand and right-hand graph for wear-out for juggling. Does it follow the same pattern? What can be learned by the wear-out phase?
- Take what has been learned by the two phases and change the process. What is the new process mean and standard deviation?

Bibliography

Agustiady, Tina Kanti; Cudney, Elizabeth A., *Total Productive Maintenance: Strategies and Implementation Guide*. 2015, Taylor & Francis Group LLC, Boca Raton, FL.

Breyfogle, Forrest W., *Implementing Six Sigma: Smarter Solutions Using Statistical Methods*, 2nd Edition, 719–732. 2003, John Wiley & Sons, Inc., Hoboken, NJ.

Dovich, Robert A., *Reliability Statistics*. 1990, ASQ Quality Press, Milwaukee, WI.

Yang, Kai; El-Haik, Basem S., *Design for Six Sigma: A Roadmap for Product Development*, 2nd Edition. 2009, The McGraw-Hill Companies, Inc., New York (1st Edition, 2003).

chapter nine

Improving by challenging the process

The design of experiments process

Objectives

- Understand the mechanics of a designed experiment.
- Create a Box Plot to show results of a design of experiment (DOE) and choosing new settings for the juggling process.
- Perform a regression analysis to optimize a single variable as a follow-up to the DOE.

This will almost certainly happen—reaching a state on any process that is a state of statistical control—but is not good enough. We still have a process creating an outcome that is neither profitable nor acceptable to the customer. The endless cycle of a good day followed by a bad day of production/quality/downtime does not seem to break. The employees are working hard, but it just does not change. Accounting has now budgeted in for a rework rate of 10%, and inspectors are added to the end of the line in a fruitless attempt to inspect the product before it hits the customer. Management begins to look at a capital infusion to increase output or quality—the latter usually in an attempt at adding in automatic inspection equipment or reworking stations. The cycle does not end. What next to break the gridlock?

The first documented designed experiment was done in 1747 by physician James Lind on the HMS *Salisbury*. In the first study, Dr. Lind studied the effect of diet on sailors afflicted with scurvy. In this study, Dr. Lind adjusted 12 afflicted sailors' diets with changes in the following: cider, sulfuric acid, seawater (yuk!), a mixture of garlic, mustard, horseradish and nutmeg, vinegar, and several fruits such as oranges and lemons. What is interesting is how he recognized the earliest form of controlling the variation in the experiment. His statement "were similar as I could have them" is commonly cited as evidence that he attempted to follow what is done today by controlling outside influences, thus increasing the power of experimentation by reducing the noise variables.

DOE was probably dormant until the 1920s and early 1930s when Ronald Fisher laid out the groundwork for future development in this area with two books on the subject most notably used for agriculture: *The Arrangements of Field Experiments* (1926) and *Design of Experiments* (1935). The late Dr. George Box of the University of Wisconsin and Dr. George Montgomery of the University of Arizona are best known today for teaching students this science and spreading the concepts from the agricultural area to the industry and eventually to Six Sigma.

How to improve any process including juggling—The juggling DOE

There are multiple methods of teaching DOE: paper helicopter and catapult are two common methods. The former will be used extensively in the next section (Chapter eleven) to demonstrate the hypothesis experimental method, ANOVA, and an advanced DOE technique. But those techniques have a common limitation—the "Interest" factor, reviewed previously in Chapter one, has its limits. The advantage of juggling is that the positive effects of the DOE are likely to remain with the student for years—and the technique to learn the improvement will be easily recalled for years versus the other methods. Why? Juggling will stay with them possibly for a lifetime, whereas the paper helicopter and catapult will have limited use after the classroom exercise.

As with the Six Sigma, the improvement effort starts after control is demonstrated in the second stage of the bathtub curve and, as pointed out by Dr. Deming, there must be a control chart on the process. Signals of lack of control that include state 1 (infant mortality) and state 3 (wear out), as shown in Chapter eight, misrepresent the concept of control. So let's take the juggling process in the middle section as an example of how to improve any process (Figure 9.1).

Is the process meeting customer specification? Let's say the customer, who in this case is probably the individual learning to juggle, would like to showcase proficiency at a minimum of seven tosses to drop. The student would gladly take more than that, but less than that would be a problem. As was done in prior chapters, the CPl or capability to the lower specification can be computed as follows from the graph given in Figure 9.1. We will not compute the CPu, or capability to the upper specification, as this, for all practical purposes, is not a factor since no upper specification exists for this process.

$$\text{CPl} = \frac{(\text{Mean} - \text{Lower specification})}{3 \times \text{Sigma}} = \frac{10 - 7}{3 \times 2.5} = 0.4$$

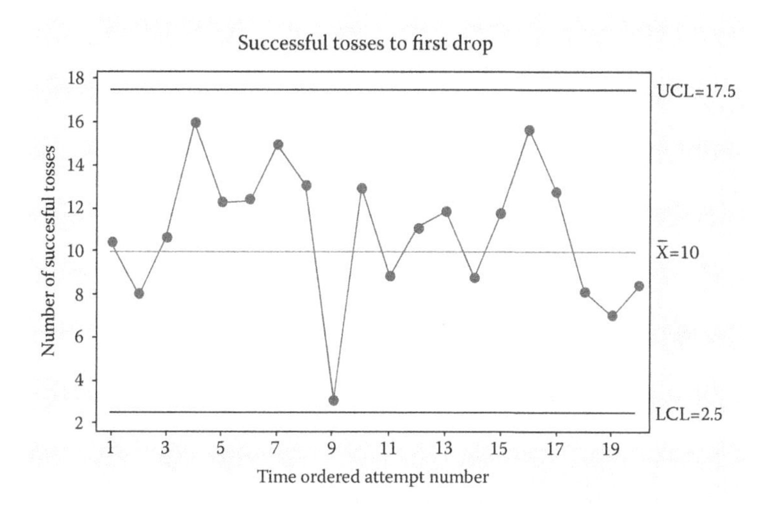

Figure 9.1 Control chart of the process in control but not satisfactory.

The rule of thumb for most processes for any CPl, CPu, or CPk is as follows:

Less than .7—major trouble
0.7–1—Trouble
1–1.33—fair to good
1.33–1.67—good to excellent
Greater than 1.67—approaching Six Sigma levels

With a CPl of 0.4, the process is clearly not acceptable to the lower specification of 7. The next step is to change the process but only after the DOE points it in the direction of positive change by reduction in variation, increase in mean, or ideally both.

The juggling DOE design

Take the process and brainstorm for every potential cause of variation. This can be done by traditional fishbone diagram or other traditional methods. At the completion of the brainstorming session, break the categories into three groups: those to be tested, those to control (not moved) during the experiment, and those we are not able to control but should be monitored. For this very simple DOE, the typical breakdown might be similar to what is shown in Table 9.1.

This is only a partial list. But the concept is one that closely follows any manufacturing or other process development phase.

Table 9.1 The brainstorming session for causes of variation

Tested	Controlled	Monitored
Ball weight	Clothing	Humidity
Height of throw	Ball shape	Ambient temperature
Wind speed	Starting hand	Warm-up and fatigue (wear out)
	Sleep habits	
	Diet	
	Warm-up and fatigue conditions	

The preliminary design

The following quote from Dr. George Box summarizes this area of the designed experiment:

> To find out what happens when a process changes, sometimes you have to change it
>
> **Dr. George Box**
> *Statistics for Experimenters*

What will be performed is a three-factor, two-level designed experiment. This is likely the most common experiment in the DOE methodology that, if done properly, will either locate an optimum setting or induce most students to pursue a higher-order DOE, which will be covered in a later section. For this exercise, each factor to be tested will be tested at two levels. Those levels will be given the code of +1 and –1. The total number of runs is eight, which can be calculated by using the following formula for three factors restricted to two levels:

$$(\#\text{of levels})^{(\#\text{of factors})} = 2^3 = 8 \text{ runs}$$

After the brainstorming session, the two levels are selected to move the process such that if the factor is significant, it will be found but not moved so far as to cause a system breakdown. The goal is to choose low and high values for the levels that are within the abilities of the juggler but have a large enough deviation to allow us to detect if they are a significant contribution to success. Examples for this process that might violate the abovementioned guideline are as follows:

- Failure of an excess weight such as a bowling ball would likely tell us nothing. Whereas a difference of 0.5 g between the low and high level would likely not find a significant difference.

- Height of toss on the high end of 1 m will likely result in all drops. The information gained by this will be obvious and irrelevant.
- For the high end of wind speed, performing in a wind tunnel with speeds above 20 kmph will result in all drops. Again, the information gained will be somewhat useless.

This is the most critical section of any DOE! A full understanding of the capability of the process is recommended before determining the levels to move the process. This is not an exact science and best developed with all operators and subject matter experts in the same room! For this exercise, the levels to be tested that were agreed upon are shown in Tables 9.2 and 9.3.

The changes in variables can be thought of as physical locations to create a visual representation of the experiment. Each factor is an axis (x = ball weight, y = toss height, z = air speed), and each level is a value along that axis. Using two levels (low and high) for each factor/axis, we can create a cube plot to represent the experiment as shown in Figure 9.2. Each of the eight corners of the cube represents a different test condition, and we will obtain data for each corner during the eight runs.

Table 9.2 Those factors to move and the test levels

Factor	Low level (–1)	High level (+1)
Ball weight (g)	7	14
Height of toss above head (cm)	15	25
Wind speed (by external fan) (kmph)	0 (off)	8 (on)

Table 9.3 The preliminary design for the variables (factors) to be tested in a typical 2-level, 3-factor design

Run	Factor 1—ball weight	Ball height	Wind speed (fan on/fan off)
1	(–) 7	(–) 15	(–) 0
2	(+) 14	(–) 15	(–) 0
3	(–) 7	(+) 25	(–) 0
4	(+) 14	(+) 25	(–) 0
5	(–) 7	(–) 15	(+) 8
6	(+) 14	(–) 15	(+) 8
7	(–) 7	(+) 25	(+) 8
8	(+) 14	(+) 25	(+) 8

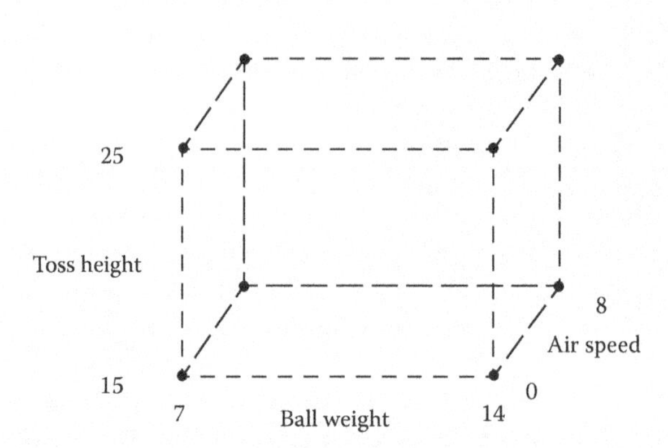

Figure 9.2 The cube plot representing the design.

Reducing the variation in those factors not to be tested

For those factors not to be tested but to be held constant, how best to accomplish this without resulting in a major delay in the DOE?

From the original brainstorming list:

Clothing: Comfortable but without areas that may interfere with the juggling. Comfortable shoes as the base is critical to the success of juggling.

Ball shape: This seems obvious, but there are options to go with odd-shaped balls. This can be a follow-up test later, but for now keep them circular.

Starting hand: This can be a factor to be moved in the experiment, as this can affect the outcome. For this exercise, keep it constant at their dominant hand starting—left if left handed, right if right handed.

Sleep habits: Similar to other athletic processes, sleep habits can be a critical factor. Under no circumstances run this on a Monday morning or following a Holiday, as the outcome may not be representative of the entire process.

Diet: Critical to the entire process. For the test, hold the calorie input constant for the day. As with all athletic events, carb up 1 h before the test. If this test is run over 2 h, add a carbohydrate-based snack between each experimental run.

Warm-up and fatigue: Similar to other process developments, this is critical to prevent but hard to monitor. With this process, best to not do this experiment at the start or end of the day as these are commonly associated with warm-up and fatigue. If a drop-off in tosses to drop occurs during the last one to two runs only, it is a signal of potential fatigue and can be confounded with a two-way interaction term. If there is a drop-off with the first two runs, it may be a sign of infant mortality or warm-up.

Environmental temperature and humidity: Best to monitor this during the testing process if it cannot be controlled. Similar to running a test on a manufacturing or nonmanufacturing process, minimize the movement in the factor if possible. This can best be accomplished by external control or minimum time for the entire DOE (such as completing within 1 day), thus minimizing the movement of such factors.

The concept of confounding variables

The concept of confounding variables is used throughout DOEs. Chapter thirteen will be an advanced DOE that will further explain this concept in more detail. But let's explore this concept with an elementary view of a confounded variable.

Look at the design and factors in Table 9.3, specifically at the third experimental factor, wind speed. Notice that the first four are at a low level (zero) and the last four are at a high level (eight or the fan on). Let's also add that the experiment is run in the same order as designed—not recommended—but without randomization of runs. What if, as a result of fatigue during the test, the diet was changed and an energy drink was taken after run 4? As a consequence, the number of tosses to failure jumped for runs 5–8. Was it the fan speed or the energy drink? That is the concept behind confounding variables, and it can happen without knowledge in many designed experiments.

So, how to prevent it? There are two primary ways. The first is to recognize it and control the extra variable; in this case, the energy drink. As was mentioned earlier, take an energy drink at the start of every run or eliminate it completely. The second is to randomize the runs. A randomization methodology will be covered in Chapter thirteen.

Sample size and randomization

How many times should we attempt to juggle during each run of the experiment? The number of data points collected is the sample size. Sample size is a function of three factors:

- The effect the input variable will have on the output. The more likely the variable to be moved will likely move the output, the less the sample size. Let's take a very basic example—the ball weight.

If changing the weight from 7 to 14 g results in the tosses to failure changing from an average of seven tosses to failure with a standard deviation of 2.5 to 50 with a standard deviation of 25, the sample size will be much smaller than a change of seven tosses to failure to eight tosses to failure. The greater the separation, the lesser the sample size. The lesser the separation, the greater the sample size.

- How much variable is the entire experiment? The greater the variability, the greater the sample size to find a difference. This is the reason behind blocking all nontested variables and a big reason why DOEs fail. The process had too much variation to "find" the difference beyond reasonable doubt.
- How much risk of an error the experimental designer is willing to take? There are two errors that must be controlled—false-positive and false-negative. False-positive is calling a process change when in fact there was none. False-negative is failing to find a difference when there was one. This risk level depends on whoever is running the experiment and the product under testing. Material for heart valves hopefully has a difference in risk level than removing material from a paper clip.

Most of the aforementioned factors cannot be fully determined until the DOE has been completed. In addition to the three aforementioned process-related factors, there are two other factors in determining the sample size: the cost of running the test and the length of time to run it. The higher the cost of each run, the fewer the runs probably going to be approved by management. The more factors to be tested, the longer the experiment and the higher the risk of adding in an unknown confounded factor (for this DOE, environmental temperature change or fatigue).

For the purposes of this experiment, we will perform seven attempts during each run.

To reduce the effect of confounding factors, we will perform the runs in a random order. Why is randomization critical, and what is the best way to randomize the runs? What happens if the DOE is run in the same order as designed and the wear-out cycle takes effect somewhat around run 5 and continues through the entire experiment? That fatigue factor is now confounded (or confused as many call it) with the third factor, the fan on and off. This confusion may result in wrongly identifying the fan or air flow as one of the critical factors.

So how to randomize? First, how not to randomize—randomize by manually choosing runs. Why not? This is a question to most instructors and students—where did you park at the university/work/home/shopping mall during the last five trips? The usual answer is the same—in the same spot. Another hands-on exercise to show how not to randomize—take four students and, without any guidance, have them stand inside a defined 2-m diameter circle. Notice what happens. By nature, they will stand equidistant from each other—not in a random pattern.

Other methods of randomization of run order may be more effective—a method utilized might be as follows: From a standard deck of cards, remove the cards ace, 2, 3, 4, 5, 6, 7, and 8. Shuffle those cards five to seven times. The card number corresponds to the run number.

The how-tos for running the experiment, and the outcome

Probably the most critical factor in the success or failure of any DOE, aside from the initial design, is the organization of performing each run. Again, Dr. George Box probably said it best, "a well-designed and ran DOE will analyze itself."

What is meant by that? Identify the critical three to eight factors and move those factors enough to see a difference without catastrophe, keeping the results well organized and the data results obvious.

What can be learned from a typical designed experiment

There are multiple possible findings in a well-run designed experiment. Three of the most critical areas are as follows:

- The factors that are behaving by themselves are commonly called main effects. These factors are relatively easy to follow-up with as they can move independent of the other variables in the experiment.
- The interaction factors. These are factors that behave differently depending on the level of a second, third, or even in some rare occurrences, a fourth variable. These are typically only found through DOEs, as will be highlighted by this experiment. When thinking of an interaction, think of the internal combustion engine—a four-way interaction between heat, pressure, carbon (fuel), and oxygen.
- The factors that have no effect on the outcome. These factors may have been identified as potential factors but with no evidence to support the findings. These are common "gut feel" factors that are determined to be nonfactors in this process. (Note: usually translated into cost reduction opportunities.)

The juggling DOE results

(Figures 9.3 through 9.5).

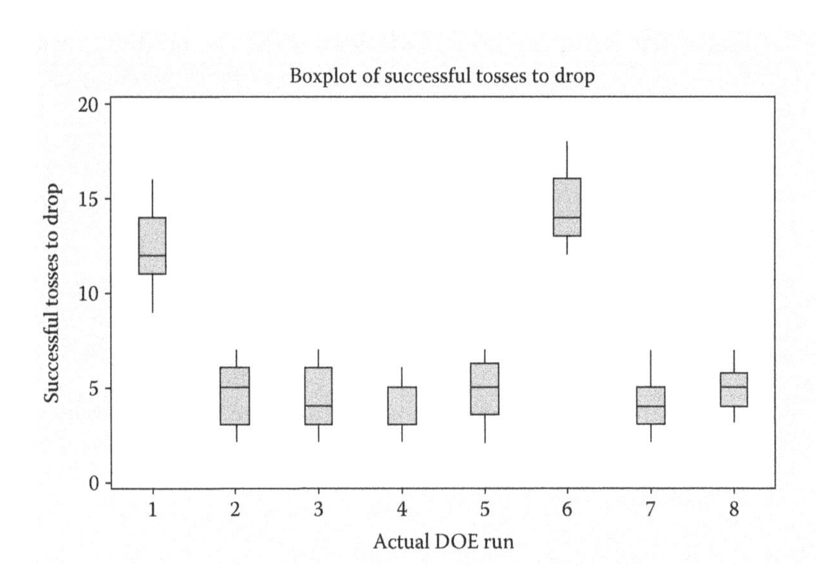

Figure 9.3 Time-ordered box plot of the outcome by run.

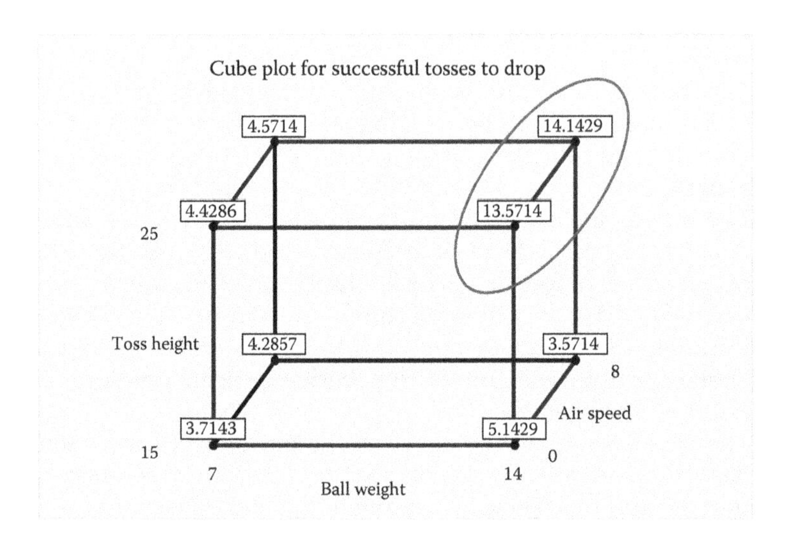

Figure 9.4 The outcome in cube plot form—interaction at ball weight 14 and toss height.

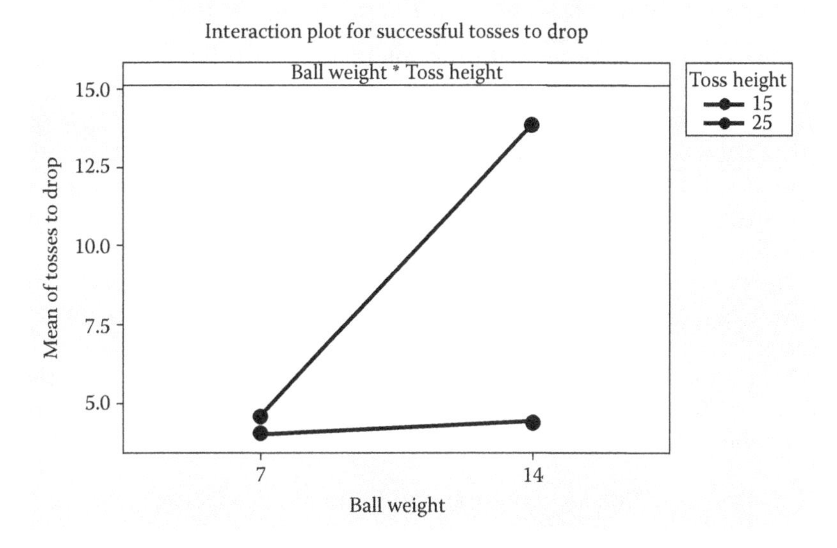

Figure 9.5 The interaction plot clearly showing the two-way interaction.

Interpretation of the results

The results clearly show two distinct areas of high performance, highlighted in Figure 9.4. There is an interaction between the ball weight and toss height. This effect clearly shows that with the heavier ball and at the greater height, the performance is higher. However, the two levels must be used together to see this improvement. Simply increasing the ball weight without changing the toss height does not result in improvement. Imagine if we had not used the DOE and instead simply changed one variable at a time to improve this process. Would we have identified this improvement? Likely not. The DOE allows us to determine when factors interact with one another and best identify the areas where the process can be improved. This also shows the other areas that have a negative effect on the performance.

There is a third finding that is commonly missed or ignored—the fan or wind effect was found to be benign or insignificant in the range of the experiment. That is a critical concept, because what was originally thought to be a factor was not—just likely gut feel. This is not unlike what is found in most industrial or nonmanufacturing operations—the data from the DOE will commonly find major opportunities for improvement and also counter the common misconceptions or historical bad judgments about what influences the process.

So what next? How best to proceed once the DOE is complete?

The opportunity the next day and the follow-up testing

The improved state found with the interaction term is critical. What is the new capability if the improvement was stopped here? Is there opportunity for improvement beyond the increased weight, increased height?

Let's answer those questions under the assumption that it is now a future day. Can the successful settings in the process be used to repeat our results? If not, how close is the process to what was found during the DOE?

The next-day results via a control chart replicating the process with the new height and heavier balls in control are typically as shown in Figure 9.6.

Notice the increase in the mean and reduction in standard deviation, which is quite common after a successful DOE is completed. New capability calculations are as follows:

$$\text{CPl} = \frac{(\text{Mean} - \text{Lower specification})}{3 \times \text{Sigma}} = \frac{13-7}{3 \times 1.5} = 1.33$$

$$\text{CPl} = 1.33$$

Typically, this would be considered good for customer performance.

We have achieved a significant improvement, but the settings may still not be optimized. To work on the next step, let's explore the ball height factor further.

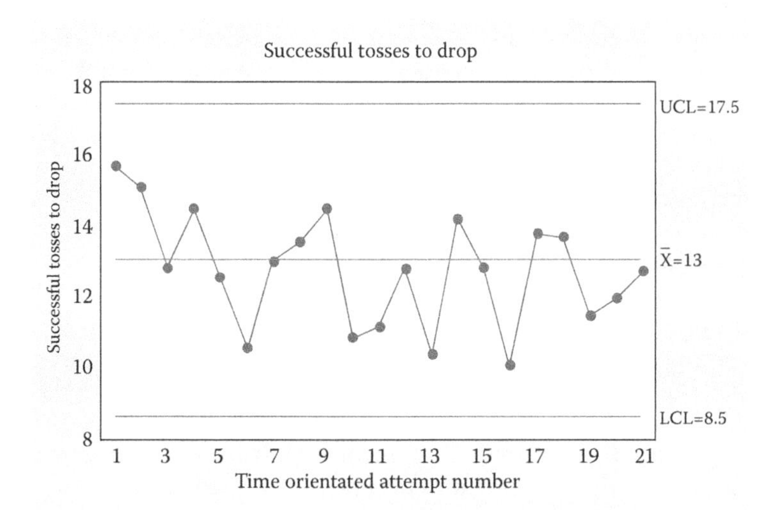

Figure 9.6 Control chart of process with greater height and heavier balls.

Next phase of improvement and the regression model

The new CPl looks impressive, but it may not be the best location of the variables in the experiment. Let's do a very simple follow-up to the model. Although there are multiple weights and shapes to juggling balls, let's lock it at the high end and see what happens when the height is changed. Is the height of 12 in. optimum? Maybe 9 or 14 in.?

With the ball weight held constant and the wind speed at the setting with minimum cost (fan off) since it was benign, let's change the height according to the data given in Table 9.4.

We can now combine results from these runs with the prior two runs from the first DOE with the ball weight at 14 g and the fan speed held constant at 0. The grand total will be five runs to effectively draw the scatterplot and make an estimate of the slope, *Y*-intercept, and regression equation (Figure 9.7).

Table 9.4 Follow-up run designs

Run	Height (cm)
9	17
10	19
11	22

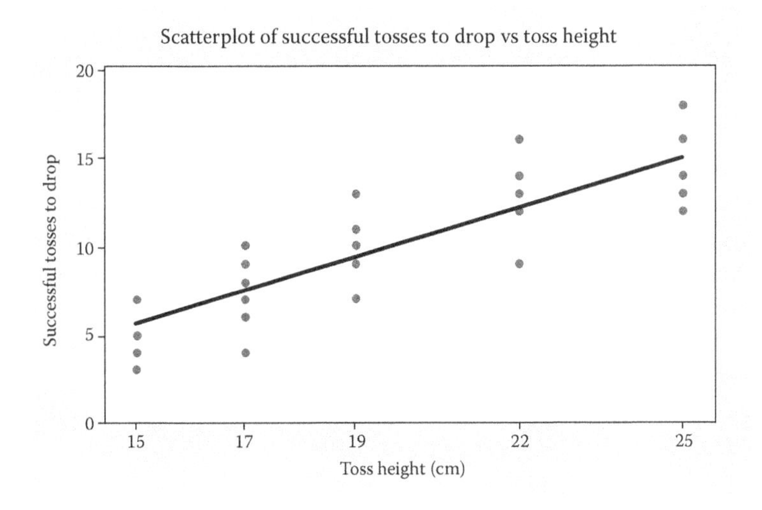

Figure 9.7 Result with sample size of seven for runs 9, 10, 11, and prior runs from the DOE.

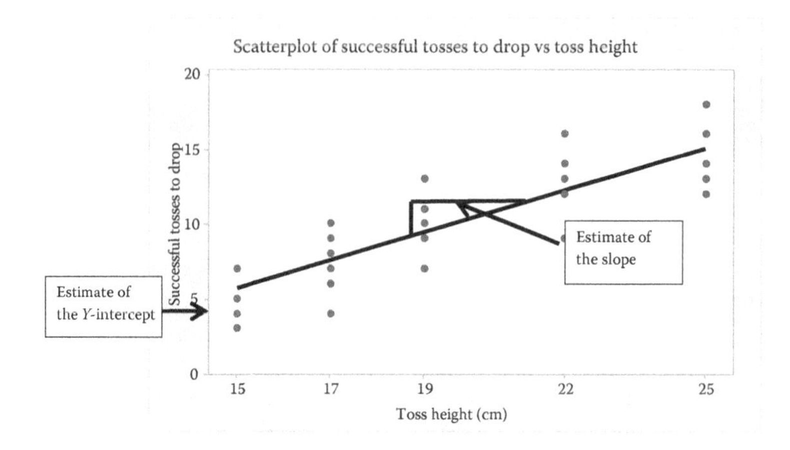

Figure 9.8 The simple mechanics of the regression model.

A very simple regression model

The regression model can be estimated by two numbers, the Y-intercept and the slope, if just looking at the single factor of the ball height. From the graph, this can be estimated by hand or by using a conventional software package.

$$Y = (Y\text{-intercept}) + (\text{slope}) \times X$$

The resulting equation with just the ball height (keeping the ball weight as a constant) is from any traditional statistical software package (Figure 9.8):

$$Y = -8.27 + 0.932(\text{ball height})$$

From this, the model can be used to find the maximum point. For this model, the height is a straight line relationship within the range under study. Thus, the maximum point would be at the height of 25. Caution with interpolating >25 as the assumption of linearity will likely not hold (the line eventually drops as the weight begins to get too heavy to handle). Although this shows an estimate of the slope, it fails to take into account a potential change in variability in tosses to drop. This will be reserved for a future discussion, but each optimization point should also include a test of variation differences.

Summary on DOE and regression model method via juggling

This process can be a challenge, but it is also an excellent analogy to implementing continuous improvement for any manufacturing or nonmanufacturing process. The improvement methodology can be applied to most any manufacturing or nonmanufacturing process. If applied in practice, the system is very similar to the steps followed. For all students learning this methodology, the interest via the juggling analogy will potentially allow them to keep this a permanent memory for years, because most will follow up with their own testing and trials outside the classroom.

Further exercises for students

- What happens to the number of runs in the DOE if the number of variables are expanded to 4? What about 5 or 6? What is gained?
- Sketch by hand the cube plot for 4–5 or 6 variables.
- What happens if, the following day, the process was lower than expected? What would you do as designer?
- Adding height and weight to the juggling process will likely add to the fatigue factor. How would you measure this effect on the new process? What would be the countermeasure?
- What about the shape of the balls? Is this potentially a factor?
- What is the optimum point for this process? Is it the maximum value? What happens to the CPk as the height is changed?
- The fan speed was tested over a limited range restricted by the fan itself. Test that theory for all students. What about 10 CFM? 20 CFM? Is it linear?
- Develop the regression model as a multiple regression with the ball weight included. What does it tell you about the interaction term?
- What if you decided that two levels were inadequate and wanted to change the design to three levels? How many runs would there be if there were three factors at three levels? What are the problems with this type of model? What method could be utilized to reduce the number of runs but still address this concern that it only looks at two levels?

Bibliography

Anderson, Virgil L.; McLean, Robert A., *Design of Experiments: A Realistic Approach*. 1974, Marcel Dekker, Inc., New York.

Box, George E.P.; Stuart Hunter, J.; Hunter, William G., *Statistics for Experimenters: Design, Innovation, and Discovery*, 2nd Edition. 2005, John Wiley & Sons, Hoboken, NJ.

Moen, Ronald D.; Nolan, Thomas W.; Provost, Lloyd P., *Improving Quality through Planned Experimentation*. 1991, McGraw-Hill, Boston, MA.

Montgomery, Douglas C., *Design and Analysis of Experiments*, 5th Edition. 2001, John Wiley & Sons, New York.

chapter ten

Design of experiments via a card trick

Objectives

- Review how design of experiments (DOEs) are designed.
- Perform a card trick that utilizes the base design of a DOE.
- Understand how this simple card trick can be utilized to solidify students' learning of DOEs.

Introduction and the card trick "mysterious mind read"

This chapter is unique in that it ties in the concept of DOE with a very unusual card trick. The concept utilizes a 5-factor, 2-level, full-factorial design in conjunction with a de Bruijn sequence to perform a mind-reading trick used with mid-to-large-size classrooms or audiences (>10 and as many as 1000). It is relatively easy to perform but somewhat difficult to set up. For all students or instructors who work through this concept, there is little doubt that the base level DOE concept will never be forgotten—especially if it is done with their own performance of the trick.

For those who believe this step may not be necessary, the author recommends proceeding to Chapter twelve, as this trick is difficult to master. Continuing in this chapter will have the added benefit of explaining the mathematics behind binary code and de Bruijn sequencing that is used in advanced mathematics and computer logic. This can also be used to strengthen the probability section reviewed in Chapters two and three.

The effect, or what the audience sees when performing for a large audience (+20)

The instructor or performer of the card trick removes what is believed to be a full deck of cards from the deck—in fact, it is a deck of 32 cards, not 52. The reason will be explained in the next section. The pack of cards is tied together with a rubber band. The deck is thrown to any audience member, who in turn is asked to throw it to a second audience member. This step can be performed as many times as desired to ensure that the audience member who receives the deck is not prearranged or known by the instructor or performer. At the completion of the fourth or fifth throw, have the audience member or student holding the cards remove the deck from the rubber band and perform a standard cut of the deck. Pass the deck to another person who cuts the deck a second time. Repeat the deck cut as many times as desired (as will be shown, the deck cut is not critical to the outcome). Pass the deck to the first person who cut the deck or any other random person, and have him or her take the top card without showing it to you. Pass the deck to a second person and have them take the next card on top. Repeat this step for five people—but only five people! At the end of the step, the deck should have been passed to multiple numbers of participants, the deck cut multiple times, and the top five cards removed of which you know the order in which the audience members removed cards from the top. Knowing the order is critical.

Inform the participants that you will be reading their minds. State, "Please look at me closely such that I can get a reading on your card." Continue with, "I am getting mixed signals, could you please help me out—all of you who have a black card please sit down." At that time, you look at each participant individually and tell them the exact card they are holding.

How and why it works

The key to the trick is the statement, "I am getting mixed signals, could you please help me out—all of you who have a black card please sit down." This tells the order of the black card/red card arrangement. For five cards, from Chapter two, there are exactly 2^5 possible combinations, or 32. Thus, the reason for only 32 cards and not 52. Every combination of the five cards has one and only one combination of the black/red card combination. Just as in Chapter nine, the 3-factor, 2-level DOE had eight unique combinations of each factor; this trick is an ideal methodology for teaching how to develop the design for the 5-factor, 2-level, 32-run design. Let's take a very simple example to show how this concept works.

The 5-factor, 2-level design as compared to the mysterious mind read

Let's look at the comparison between the mysterious mind read and the 5-factor, 2-level, full-factorial DOE as shown in Table 10.1. Similar to the 3-factor, 2-level DOE from chapter nine, this five factor DOE design is shown in Table 10.1.

The design shown in Table 10.1 is a standard design, if in the rare instance, a 5-factor full-factorial DOE was implemented (more on that in Chapter twelve). It is unusual to have the resource and time available for

Table 10.1 5-factor, 2-level, 32-run full-factorial design

Factor A	Factor B	Factor C	Factor D	Factor E	Factor A	Factor B	Factor C	Factor D	Factor E
–1	–1	–1	–1	–1	–1	1	–1	–1	–1
–1	–1	–1	–1	1	–1	1	–1	–1	1
–1	–1	–1	1	–1	–1	1	–1	1	–1
–1	–1	–1	1	1	–1	1	–1	1	1
–1	–1	1	–1	–1	–1	1	1	–1	–1
–1	–1	1	–1	1	–1	1	1	–1	1
–1	–1	1	1	–1	–1	1	1	1	–1
–1	–1	1	1	1	–1	1	1	1	1
1	–1	–1	–1	–1	1	1	–1	–1	–1
1	–1	–1	–1	1	1	1	–1	–1	1
1	–1	–1	1	–1	1	1	–1	1	–1
1	–1	–1	1	1	1	1	–1	1	1
1	–1	1	–1	–1	1	1	1	–1	–1
1	–1	1	–1	1	1	1	1	–1	1
1	–1	1	1	–1	1	1	1	1	–1
1	–1	1	1	1	1	1	1	1	1

the earlier design. A much more efficient design will be shown in Chapter twelve. But for now, the +1 and –1 represent the high and low levels of each factor. In this design, all combinations are covered. The cube plot for this design would look like that shown in Figure 10.1.

Now compare that design with the setup for the mysterious mind-reading trick. The trick is set up exactly to the earlier design with a de Bruijn sequence, which will be covered later (Figure 10.2).

Notice the similarities with the original 32-run DOE. Every combination of red and black cards can be found in the earlier set design, and every combination is unique—they do not repeat. If we all agree that each row is unique, how does the order of the 32 cards relate to the earlier? Welcome to the world of the de Bruijn sequence.

There are two methods for setting up this trick. We will cover both in the following section as they both can be used for educational purposes. The first method is more for understanding the design, and the second is for understanding de Bruijn sequencing and binary code.

Method number one for setting up the card trick: The easy method

The easiest method is to assign the cards by the following sequence. Cut and paste the format as seen fit as it follows a de Bruijn sequence, which will be covered in detail later. The 32-card sequence is as follows:

8D, AD, 2D, 4D, AH, 2S, 5D, 3H, 6S, 4H, AC, 3S, 7D, 7H, 7C, 6C, 4C, 8C, AS, 3D, 6D, 5H, 3C, 7S, 6H, 5C, 2C, 5S, 2H, 4S, 8H, 8S

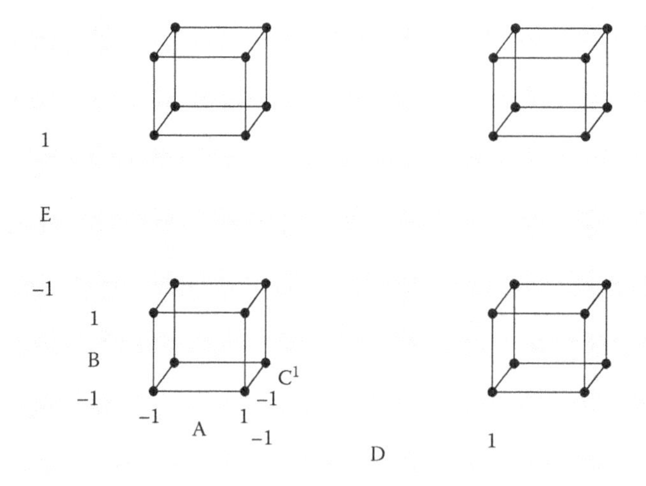

Figure 10.1 5-factor, 2-level, full-factorial cube plot.

Figure 10.2 Setup for the mysterious mind-reading card trick.

The cutting of the deck a multiple number of times changes only the starting point of the sequence—it does not change the order. Thus, the magic of this trick. The first person who pulls the card from the top determines the starting point—the next four cards removed from the top of the deck follow in sequence. Each possible five-card sequence has one and only one starting point. Notice the cards are ace through 8. All cards from 9 through king have been removed. As an option, to ensure the mystery of the trick, the king can be substituted for either the 6 or the 9, the jack for the ace, and the queen for the 8. The determination of which cards to switch is strictly independent. Rarely will the audience be suspicious, since they only see the top five cards.

Method number two: An advance method by binary code and de Bruijn sequence

The sequence developed earlier is a method used in computer science and commonly known as a de Bruijn sequence, named after Dutch mathematician Nicolaas Govert de Bruijn.

Let's start with reviewing the binary sequence of mathematics. Our conventional method of mathematics is base 10. The number 12 is $10 \times 1 + 2 \times 1$. The number 594 is $5 \times 100 + 9 \times 10 + 4 \times 1$. Binary works in powers of 2 rather than of 10. The sequence 111 represents $1 \times 4 + 1 \times 2 + 1 \times$

1 = 7. Notice that the multipliers 1, 2, and 4 are now multipliers and not base numbers. Using this formulation, the card numbers of ace through 7 can be represented by the following:

Ace or 1	001 $(0 \times 4 + 0 \times 2 + 1 \times 1)$
2	010 $(0 \times 4 + 1 \times 2 + 0 \times 1)$
3	011
4	100
5	101
6	110
7	111

For this card trick, only aces through 8s are utilized, which leaves only the card of 8 without a designation. For this trick, utilize 000 as the number 8.

For the suit designation, utilize a two-number code. The following is a standard code but can be modified as needed:

Diamond	00
Heart	01
Spade	10
Club	11

The five-digit code for all cards: Aces through 8—In binary code

The first two digits represent the suit code of the card, and the last three, the three-digit binary code for the number.

Using this method, the five-digit number 01001 is the ace of hearts. How was that derived? The first two digits, 01, is the heart code; the last three, 001, is the binary code for 1 (or ace). By that same method, 11010 is the two of clubs—11 represents the suit clubs; the 010 represents the number 2.

Another concept used during DOE is the table of contrast development used in this next sequence. In computer science, this method is called adding by "modulo two." The adding sequence is as follows:

$0 + 0 = 0$
$0 + 1 = 1$
$1 + 0 = 1$
$1 + 1 = 0.$

Start with the ace of diamonds, which will be only one of two cards to remember for the sequence and rethinking in binary code. Ace of diamonds: 00001. The other card is the 8 of spades (10000), which will be covered next. Start with the ace of diamonds and add by the modulo two sequence the first and third numbers in the card number sequence. Add that onto the chain and continue that pattern until 31 cards are completed.

The exercise is as follows:

00001 Ace of diamonds

By modulo two method described earlier, add the first (0) and third (0) digit and place the result (0) at the end. The code for the second card begins with the second digit in the sequence; the second card is 00010 or the 2 of diamonds.

000010 Ace of diamonds (00001) followed by the 2 of diamonds (00010)

Add first (0) and third (0) digit from the second card and place result (0) at the end. The third card is 00100, the 4 of diamonds.

0000100 Ace of diamonds, 2 diamonds, 4 diamonds

Add first (0) and third (1) digit from the latest card and place the result (1) at the end. The fourth card is 01001, ace of hearts.

00001001 Ace of diamonds, 2 diamonds, 4 diamonds, ace of hearts

Continuing on in that sequence results in the following 31-card sequence:

0000100101100111110001101110101

This 31-card sequence is missing the 8 of diamonds or 00000. Place an additional zero at the start of the sequence, and that will be matched up with the sequence of 8D, AD, 2D, 4D, AH, 2S, 5D, 3H, 6S, 4H, AC, 3S, 7D, 7H, 7C, 6C, 4C, 8C, AS, 3D, 6D, 5H, 3C, 7S, 6H, 5C, 2C, 5S, 2H, 4S, 8H, 8S identified earlier.

For an advanced methodology, to be able to name the five cards with just the card's black/red designated and not the exact card numbering, recall the patterns in the full set of card sequences in Figure 10.2. Next, recall the sequence of the first card in each group of 8. The first card is in the order of 8-ace-2-3-4-5-6-7 by the four suits (again, refer to Figure 10.2). Knowledge of the first card in the sequence will tell you the remaining four cards by applying the above de Bruijn sequence for the next four cards. This will have the effect of not needing to know the exact card order to perform this trick. This is very difficult to master but very effective.

How it works—The effect

So very basically, if the five participants stand up or sit down according to the following sequence of the cards

Person 1	Person 2	Person 3	Person 4	Person 5
Sitting	Standing	Standing	Sitting	Sitting
Black	Red	Red	Black	Black

Sequence row: 3 of spades, 7 of diamonds, 7 of hearts, 7 of clubs, 6 of clubs.

Summary and potential next steps

This card trick is best performed after learning the 2-level, 3-factor, 8-run DOE and the opening probability section. It will solidify the education process for how to design most of all DOEs. Following through with the concept of every option of black and red card combinations in 5 factors has been covered. The plan would be at that time to add in other full-factorial DOEs to the mix, and if successful, the concept will significantly add to useful, relevant, and now interesting/shocking, at higher levels.

Suggested follow-up steps for classroom exercises:

- What does a 4-factor, full-factorial look like? What about a 6-factor?
- Discuss the risk associated with running a 5-factor full-factorial.
- What options are there to a 5-factor full-factorial if because of cost or time limits, it cannot be done?
- For the card trick only. What would happen if you use 52 cards instead of 32? How could you modify the trick to have the same outcome? Hint on the method: There are 20 of the 52 cards that will have duplicate corners on the cube plot or about 45%. If one of those is chosen, ask one person who has a red card if it is heart or diamond. That will eliminate one of the two points at the corner. Same applies if it is a black card. Ask if it is a spade or club. The answer will eliminate one of the points on the corners.

Author's comments

DOE is a very powerful tool for industry and transactional areas, and if taught properly, it can be learned by everyone. Having ran first-hand over 100 DOEs at multiple types of industries, the key to understanding DOE is the cube plot. This exercise transitions from an industrial experimental environment to a game-type activity. This chapter demonstrates one of the most difficult DOEs simplified for most. Most students will reach the Aha moment after this exercise, and their understanding and confidence will expand into other uses of this technique.

As an extra added effect, before the step when the students are asked what the color of the card they are holding is, ask them a series of trivial questions: Candidate number 2, what month were you born in? Candidate number 3, what is your height in centimeters? Candidate number 5, what is the first letter of your middle name? Why? They have nothing to do with the outcome? Correct, but how many product investigations are hampered by useless and nonessential data and opinions? This is an excellent demonstration of the potentially damaging outcome to those types of questions and a way to prevent them from overcoming an investigation. Nothing but the relevant facts please!

Bibliography

Ben, David, *Advantage Play: The Manager's Guide to Creative Problem Solving*. 2001, Key Porter Books, Toronto.

Box, George E.P.; Stuart Hunter, J.; Hunter, William G., *Statistics for Experimenters: Design, Innovation, and Discovery*, 2nd Edition. 2005, John Wiley & Sons, Hoboken, NJ.

Diaconis, Persi; Graham, Ron, *Magical Mathematics: The Mathematical Ideas that Animate Great Magic Tricks*. 2012, Princeton University Press, Princeton, NJ.

Hoy, David. *The Bold and Subtle Miracles of Dr. Faust*. 1963, Ireland Magic Co., Chicago, IL.

part three

Introduction: Data, statistics, and continuous improvement via the experimental helicopter

chapter eleven

Hypothesis testing method utilizing the experimental helicopter

Objectives

- Understand the mechanics of the hypothesis testing process.
- Understand why this process is critically important for any continuous improvement process.
- Learn how to teach hypothesis testing using the experimental helicopter.

A brief history of hypothesis testing

Hypothesis testing can be traced back to the 1700s, with statements made by mathematician Pierre-Simon Laplace. Likely the first documented writing on the science was with Ronald Fisher in 1933. In his paper, the concepts of the null hypothesis, rejecting the null hypothesis, and critical value were first identified. In a later paper, Jerzy Neyman and Egon Pearson, who argued heavily with Ronald Fisher, added the concept of the beta error along with other concepts utilized today.

The model has been disputed over the years, but the intent should never be in doubt: With proper understanding and implementation, the concept can make major changes in any process, while at the same time preventing the making of any harmful changes brought about by following false signals. The specific model utilized (Ronald Fisher or Neyman/Pearson) is for the most part irrelevant. This book's explanation will closely follow the Neyman/Pearson model. Understanding and applying the concept correctly at the front line of most operations, by middle management responsible for continuous improvement or even at the executive management level, is critical. The exercise utilizing the experimental helicopter will not only add to the understanding but will also help to add the third dimension to the recall function—the "Interesting/Shocking" factor.

Table 11.1 The courtroom model

	What the jury found	
What really happened	**Not guilty**	**Guilty**
He/she really did it		
He/she really did not do it		

The hypothesis testing model

Let's start with the traditional courtroom analogy for hypothesis testing and expand it with a different twist to help those not familiar with the model understand it better.

The four quadrants of the jury outcome for a trial are shown in Table 11.1. The column "what the jury found" represents the outcome of a jury trial while the column "what really happened" represents what the defendant really did. For the remainder of the explanation, we will utilize the gender-specific term for masculine. This not to say that the model analogy does not work with a female as the model works fine with Joan of Arc, but for reasons to be discussed later, we will utilize the male gender.

Before labeling the four boxes and analyzing terms, notice the language, which is similar to the language used in the hypothesis testing model—guilty or not guilty. Notice the model never states innocent. This is one reason why the courtroom model works well: Typical courts of law must determine between two outcomes, guilt beyond reasonable doubt and that there is reasonable doubt of guilt, which is not the same as innocent. That is the same for hypothesis testing—either there was sufficient evidence to support a change in the process beyond reasonable doubt or there was not sufficient evidence. At no time does the data support an argument that the processes are equal. That is not part of the courtroom model, nor is it part of the hypothesis testing model. Along with other areas to be described next, understanding that statement is critical to the continuous improvement process in any manufacturing or nonmanufacturing environment. Gut feel and emotions have no part in the hypothesis testing process—and likely helps explain why companies that stick to the model thrive in the long term.

Making the model less analytical

The next steps with the courtroom model usually solidify the concept of experimental power and the two error scenarios that will now be covered. The fourth box, which historically has little meaning, will be explained in more detail via the experimental helicopter.

The two error boxes in the model are traditionally characterized by alpha and beta errors, which can be difficult to understand for those not familiar with applied statistics.

Alpha error or type I error

This is represented by the intersection of the "guilty" column and the "he/she really did not do it" row (bottom right box of the four). This traditionally has been labeled as rejection of the null hypothesis when the null hypothesis is true. More on this later in this chapter concerning the null hypothesis. This is better understood if changed to something a nonstatistician can connect to. If this is a courtroom trial, who best represents someone who was found guilty but in reality never did it? The example utilized is that of ex South African President Nelson Mandela. Open to argument, but most scholars are in agreement that he was falsely convicted for the crime of overthrowing the state in 1962 and imprisoned for 27 years. This analogy will assist the recall of this area as Mandela is generally easier to understand than, "rejection of the null hypothesis when the null hypothesis is true." What are we basically saying? This is making the claim that there was change in the process, but the evidence or data never supported it. This was followed by actually changing the process and not receiving the expected benefits of the change.

Beta or type II error

This is represented by the intersection of the "not guilty" column and the "he/she really did it" row. This traditionally has been labeled as failure to reject the null hypothesis when the null hypothesis is false. Again, change this section to something a nonstatistician can relate to utilizing a courtroom analogy. This box can be represented by either notorious Chicago gangster Al Capone, or a more modern iconic figure, OJ. Either will work as an understanding of the concept behind beta error or type II error. In has been reported that Al Capone was a master at the process of avoiding prosecution for crimes he committed. How does this relate to the hypothesis model? This is the change that would have improved the process but was not found because of insufficient data or evidence to support the change. Consequently, the process does not improve and stays stagnant.

The Power box

This and the other box are best understood after fully understanding the two error boxes and their associated negative outcomes. The Power box is the intersection of the Guilty column again under the "what the jury found" column and the "he/she really did it" row under the "what really happened" row. This box can be represented by any criminal who really did it and was found guilty—I like to use investor-turned financial swindler Bernie Madoff as an example in this box. Criminal? No doubt about that! Found guilty? Yes. Action taken? Yes—serving a lifetime prison term. Society better for this action? No doubt about that. This box is called power and is commonly associated with a major find in the testing process. This box is also called "rejection of the null hypothesis when the null hypothesis is false." That statement is confusing to most people who do not understand applied statistics. It is much better understood by the Madoff box.

The other box

The fourth and final box is the interception of the "not guilty" column under the "what the jury found" and "he/she really did not do it" under the "what really happened" rows. This typically has not been given a statistical name nor has it been given credit as a success, which is unfortunate. As a close analogy from Chapter four, think of Sally Clark on her second trial when she was found not guilty (and really did not do it). This box has numerous huge benefits to manufacturing and nonmanufacturing operations. Findings with sound statistical fundamentals, this box can be used to disprove practices justified by instinct and/or "gut feels" which may have been alpha errors in the past. A finding that falls into this box is commonly discounted as a failure when in fact it can be used to reduce cost of operations, if the factor put under the model is truly not a factor, and a reduction in the quantity can be used to reduce cost of operation.

Transitioning from the courtroom to the production model

Once the concepts of alpha, beta, and power are understood, there are typically a few minor definition steps analogous to the courtroom before transitioning to the production and experimental helicopter model.

The null hypothesis (Ho)

The courtroom analogy to the null hypothesis is "presumed innocent until proven guilty." This one is difficult to understand for most first-time students in this area: For any process change, there is an assumption of no difference unless proven guilty beyond reasonable doubt. Not the other way around, which happens often in manufacturing: There is a perceived difference, and it must be proved there is none. In a courtroom, he/she is not assumed guilty and proven innocent. This may seem obvious, but in real manufacturing processes, this happens repeatedly. Here are a few typical examples:

- A production process had 10% scrap today and was characterized as disastrous. The operations manager wants an explanation on his desk by 5:00 pm. A further explanation of the process shows the scrap averages 7% and follows the binomial distribution. 10% scrap is an expected outcome of the process.
- An employee has an accident in the heat treat department. This was the second accident this year for the employee. The employee is given written notice and released. A further look at the data for the shop shows an average of four accidents per month with a total of 300 employees. This second accident models the Poisson distribution and shows no significant change from expected results.

Both of the above are examples of mixing special cause variation with common cause variation. Addressing the latter, common cause variation, requires changing the entire process. This might require asking what the reasons are for all the accidents over the past 2 years. Is there a common area such as location, time of day or shift, age of employee, years of experience, day of the month, etc.?

The alternative hypothesis (Ha) and significance level

The alternative hypothesis is similar to a guilty verdict in the courtroom. The significance level is very similar to "beyond reasonable doubt." Notice how the two work together: Beyond reasonable doubt must precede a guilty verdict. In any process improvement that follows this strategy, before a change is implemented, there must be supporting evidence (the data) beyond reasonable doubt. The question is, at what level is it necessary to prove guilt? Historically, this has been either 1% or 5% depending on the risk level of the prosecutor. With the advent of computer analysis, the exact level of risk is easy to calculate, and the risk level may depend on the critical nature of the product or process to be changed if the data supports guilt. The higher the risk of the process or product (such as the lining thickness in a heart valve), the higher the critical level.

The final courtroom analogy might look like that shown in Table 11.2.

Notice that the model is the basis for the statistical process control chart covered in a prior chapter. Reviewing the model, the null hypothesis is the control chart without out-of-control points. Any point outside the upper or lower control limits is analogous to "rejection of the null hypothesis and acceptance of the alternative (Ha)". In that case, before taking action, there must be evidence beyond reasonable doubt that there was

Table 11.2 The hypothesis test part 2

	What the jury found	
What really happened	Not guilty	Guilty
He/she really did it	OJ, Al Capone or Beta error	Madoff or power
He/she really did not do it		Mandela or Alpha error

a change in the process (rejecting Ho). The concept of significance is the upper and lower control limits. They are set to guide an operator to take no action until rejecting the null hypothesis at the ±3 sigma level.

The production model

The same courtroom analogy applies to the production or operations model. In this model, there are basically four areas that must be determined before utilizing the hypothesis model (Figure 11.1):

- The underlying distribution
- The significance level
- The separation distance that means something
- The process variation
- The process means under the null and alternative hypothesis test.

Notice that the sections of the theoretical model include alpha, power, the null hypothesis, and the alternative hypothesis. But more importantly, the model highlights the multiple relationships between the power and the two errors, alpha and beta. For those responsible for implementing the model in manufacturing or nonmanufacturing, this relationship is at the heart of the success or failure of any continuous improvement system. Looking at the aforementioned model, some of the following critical points to the experimentation process should be relatively clear:

- A reduction in variation in either the null hypothesis or alternative hypothesis reduces beta error and increases power. That should be obvious from the graph, but often this is not understood in manufacturing, transactional services, and other areas. Reduction in experimental variation can increase power sometimes by a factor of two or three. Examples of this will be provided later on in this chapter.
- Increasing alpha error will increase power but at a price. Alpha errors are the silent killers of the experimental process. A beta error

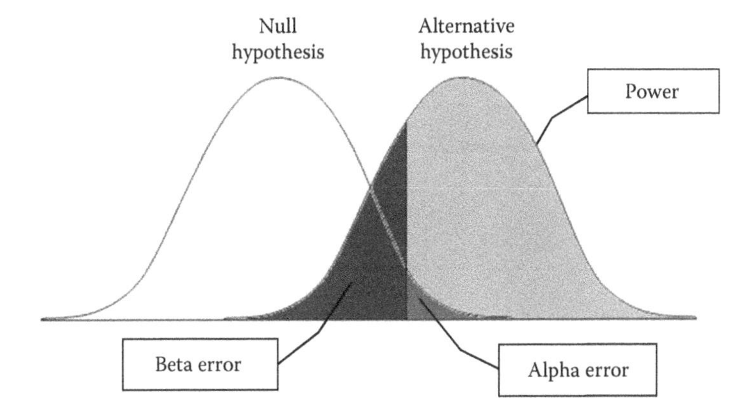

Figure 11.1 The hypothesis model.

may result in the failure to find a significant result, but the alpha error can and will result in incorrectly stating a tested variable was successful when in fact it was not. This commonly leads to a credibility factor in future testing on top of the potential negative effects of changing a factor that should have been left constant.

- The further the distance is from the mean of the null hypothesis to the alternative hypothesis, the greater the power. This point is critical for experimental testing and Design of Experiments (DOEs). Unlike the first bullet point, increasing delta may or may not be possible (whereas a reduction in variation is almost always possible). But the concept in DOE of requiring movement beyond comfort zones is driven by this concept; all areas are equal, and the greater the split between the null and the alternative, the greater the power or likelihood of a significant find.

Demonstrating the experimental process with the paper helicopter and rack system

The paper helicopter has been around for decades as a technique to demonstrate DOE. In this process, we will demonstrate how to use this for the experimental process and the negative effect of variation on the power value of experimentation. To demonstrate, we will utilize a rack system to facilitate changes in dropping height. See Figure 11.2 and Table 11.3 for descriptions of how to make the paper helicopter and how to utilize the support rack system to change the height (Figures 11.3 and 11.4).

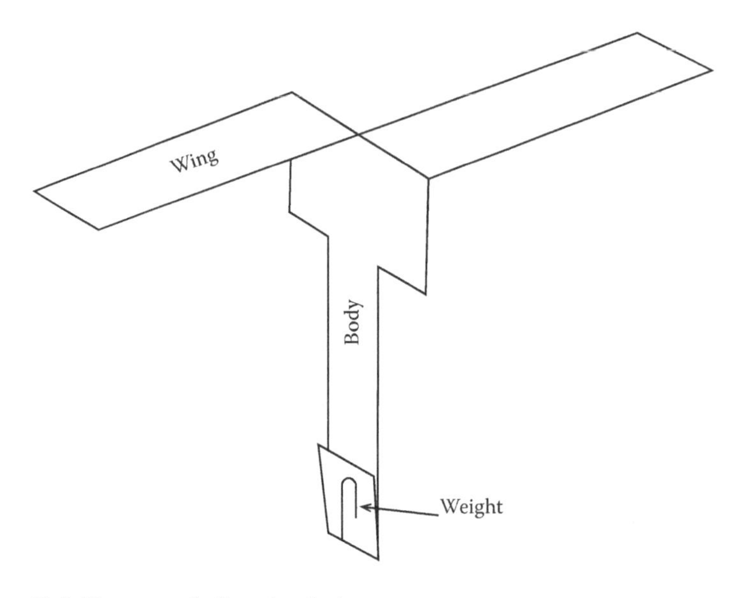

Figure 11.2 The paper helicopter design.

Table 11.3 Typical dimensions for base line construction

Helicopter characteristic	Extreme low end	Extreme high end
Wing length (mm)	70	120
Body width (mm)	35	50
Paper clips or weights (number of)	1	3
Body length (mm)	70	120
Wing width (mm)	40	60

Figure 11.3 Picture of rack with helicopter in flight.

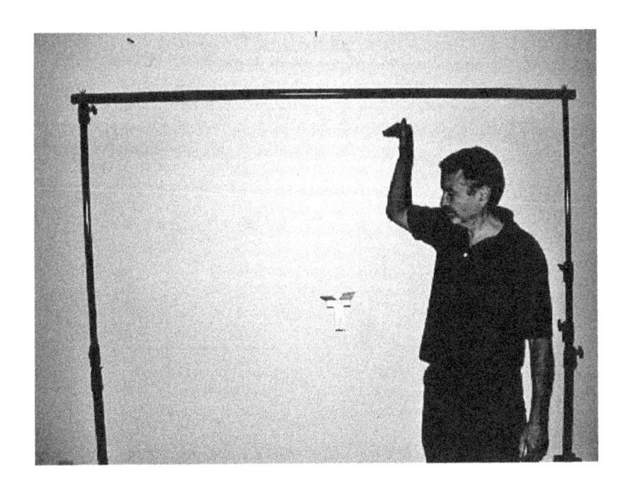

Figure 11.4 Picture of rack with helicopter in flight.

A test to demonstrate the hypothesis testing method and the negative effects from variation

Take a series of seven helicopters as if they were produced off a manufacturing line. The seven might have the dimensions shown in Table 11.4.

Set the rack up for a two-and-half-meter drop from the center of rack to the floor. Adjust the right side of the rack to 10 cm higher than the center (this is the advantage of utilizing the racks as the flight is typically dropped from the ceiling or other fixed height. This limits the ability to adjust the floor-to-height distance). Have all students either in pairs or in three- to four-person teams (the latter is better for logistics). Drop 21 flights (the seven helicopters each replicated three times) from the center and the right side. Follow the technique in Chapter four to randomize all drops. For each flight, record the time from releasing the helicopter until it reaches the floor.

The purpose of this test is to determine if there is a significant level of difference between the center and the right side. The two distributions should look similar to that shown in Figure 11.5.

This is an exercise in training about the problems of excess variation. Based on the high variations from the seven helicopters flown, the following conclusion can be drawn:

- With an alpha percentage of 5%, there is insufficient evidence to support that the right side is greater than the left side, even with the mean differences showing a >0.01 s difference in flight time. This can be determined visually or mathematically (shown at the end of the chapter).

Table 11.4 Helicopters as pulled from production line

Helicopter #	Wing length	Body width	Wing width	Body length	Paper clips
1	70	35	50	70	1
2	120	45	50	120	3
3	120	50	50	120	2
4	120	35	50	120	1
5	120	45	45	70	1
6	70	50	45	70	3
7	70	35	45	70	2

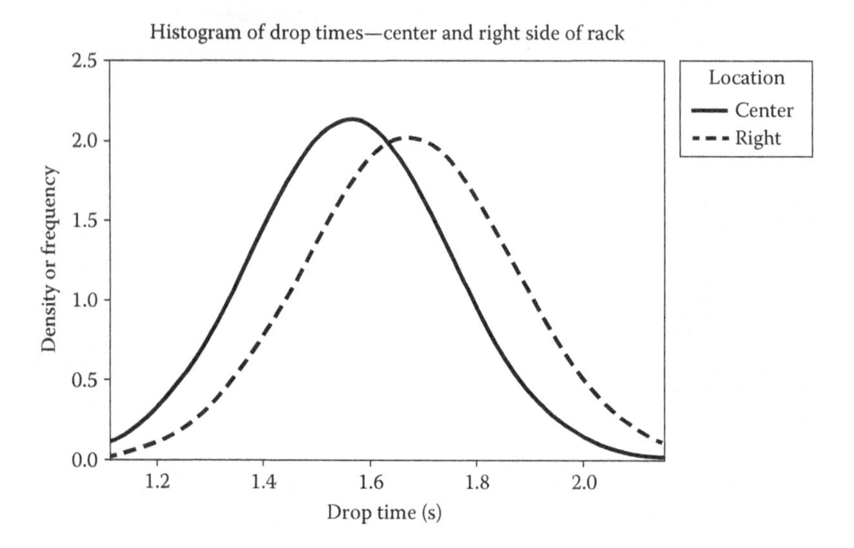

Figure 11.5 Drop times from center and right side of rack.

- There is a large beta error at ~80%; consequently, the power is ~20%. The factor likely hurting the positive test outcome is the high variation in product entry, gauge R and R, or variation in helicopter design.

This test clearly represents the problem with excessive variation. There may have been a difference in means, but because of the other factors not held constant, the opportunity for improvement will not be proven. Note that this outcome does not imply that there is no difference between the left and right side, but only that there was insufficient evidence to support that conclusion. Thus no statement could be made verifying the difference between left and right. Poorly designed and ran test—yes. But still unable to draw a conclusion based on the data. There is no opportunity for improvement unless the test is redesigned with the variation reduced, difference between the null and alternative hypotheses is increased, or there is a very large increase in the alpha error.

A better test to determine if there is difference: Hold constant the helicopter design

Excessive variation is the evil of a successful experiment. It is also one of the keys to optimization in Six Sigma. For the next test, we will demonstrate the evils of excessive variation by showing the benefit when variation is reduced. To demonstrate this, repeat the last test with the helicopter design held constant. In this design, take helicopter 3 from above: Wing length 120, body length 120, body width 50, wing width 50, weights 2. Drop and time the helicopter 21 times. The typical results are as shown in Figure 11.6.

From the graph, notice the difference in the power. Utilizing the same method as employed in Figure 11.3, with nearly the same means, the significance level on the difference is beyond 5%. Also notice the beta error is approximately 20%, which makes the power at 80%. Compare this with the earlier example, which was a test to answer the same question but leading to a different conclusion. The different conclusion in this example was only found after reducing the variation by holding the incoming parameters (the helicopter design) constant.

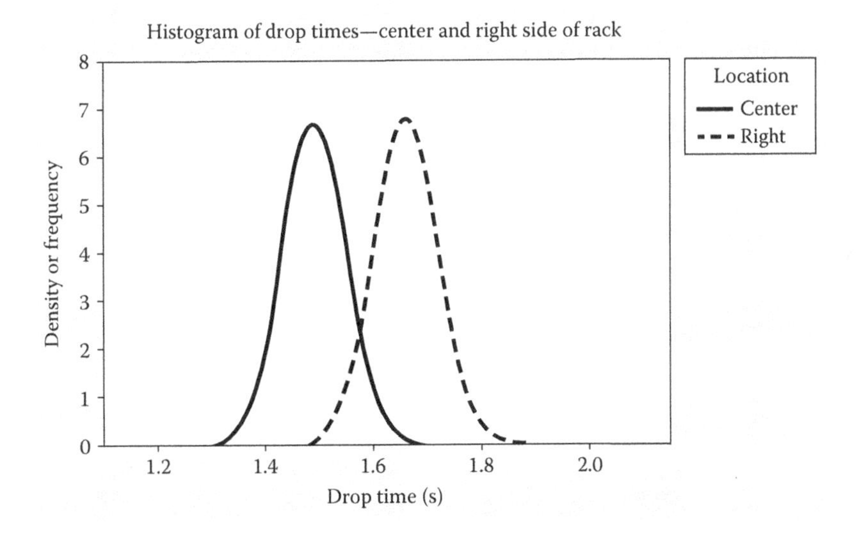

Figure 11.6 Drop times—Standard deviation reduced from 1.5 to 0.5.

Summary of the hypothesis testing method and the use of experimental helicopter

The courtroom analogy and the experimental helicopter together are a tremendous technique for increasing the "relevant, useful, and interesting/shocking" factor, leading to a much better storage and option for retrieval once completed. This methodology will commonly lead to follow-up questions regarding different facts presented in a working environment, home life, or in the news. Such questions or comments might include the following:

- We had a horrible day yesterday for scrap as it climbed to 8%. What is the confidence interval around the average of 4%? How does this lead to dispel the notion of "are you telling me the scrap rate doubled and we should not care?" What is the counterargument to that statement?
- A report came out today that showed the mean income in this country increased from 51,000 to 52,000 since 2004. What is the next question? How does your answer change based on the risk level of a wrong answer?
- Good news, the average fuel economy of a model X increased from 31.5 to 32.1. That finding will result in a 0.2% decrease in NO_x emissions over the next 20 years.

The use of hypothesis testing is an important tool to be used in any continuous improvement system. Once understood, the application will likely point to improvement opportunities in any operation. Properly utilized, the experimental methodology as shown by the hypothesis model can be the framework around a continuous improvement system.

Author's notes

Having done this method in class and in actual operations more than 100 times, the following points are critical:

- The key to understanding continuous improvement models for most operators, administrators, and leaders is this model. The alternative is the constant adjustment of any process and living with the negative side effects from adjusting processes that do not demonstrate a clear signal.
- As was stated, the paper helicopter model has been around for years but not commonly used for the hypothesis model. In conjunction with the racks, the helicopter is an effective tool for teaching this method.

- Reduction in beta error is the output of the reduction in variation. This is directly tied to the experimental power. Graphically understanding the concept behind beta and power is critical to the application. Once students plot the graph of the results from the helicopter experiment, the confidence and understanding will follow.
- The mysterious box in the model is the "no name" box or the interception between, "not guilty" and the "he/she really did not do it." This box is underrated, powerful, and full of optimization opportunities. When the author conducts experiments, variables in this box will consume more than 30% of the follow-up opportunities. These are gold mines ready to be found.

Bibliography

Fisher, Ronald, Statistical Methods and Scientific Induction, *Journal of the Royal Statistical Society, Series B*, 17 (1), 69–78, 1955.

Hayes, William L., *Statistics*, 5th Edition, 279–283. 1991, Harcourt Brace College Publishers, Fort Worth, TX.

Laplace, P., *Memoire Sur Les Probabilities, Memoirs de l'Academie Royale des Sciences de Paris*, 1778, 9, 227–332, 1781.

Lehmann, E.L., Introduction to Neyman and Pearson (1933) on the problem of the most efficient tests of statistical hypotheses. *Breakthroughs in Statistics*, Volume 1, (Eds., Kotz, S., Johnson, N.L.). 1992, Springer-Verlag, New York.

Neyman, Jerzy; Pearson, Egon S., On the Problem of the Most Efficient Tests of Statistical Hypotheses, *Philosophical Transactions of the Royal Society A*, 231 (694–706), 289–337, 1933.

chapter twelve

An intermediate use of the experimental helicopter

The 5-factor, 2-level, 16-run half-factorial designed experiment

Objectives

- Execute an intermediate design of experiment (DOE) using the experimental helicopter.
- Understand the power behind the 5-factor, 2-level, 16-run half-factorial DOE.

An overview of this model and why it is so successful at improving any process

This model (5-factor, 2-level, 16-run half-factorial DOE) has been published in other books and articles numerous times. From a practical perspective, the author has used this model over 20 times in manufacturing and non-manufacturing operations. A mere 16 runs, if used properly, can improve almost any operation by 10%–40%. Why is it so effective? Because almost all existing processes have two to four critical input parameters. That was the insight behind Dr. Joseph Juran's Pareto concept, and it is still alive today. The issue holding back most manufacturing operations is it is unknown which two to four parameters are critical or which interactions are associated with those two to four.

This chapter will describe the DOE from a practical standpoint through the use of the experimental helicopter. The data analysis will be kept to a minimum with the model results presented using a visual perspective. An optimum design method that can be done simply in a very fast and efficient method without the use of statistical software will be discussed. Although often a necessary asset, at times, especially for entry level students, the use of statistical software can be overwhelming and/or intimidating having a potential detrimental effect on the learning process. For those who are unfamiliar with the base DOE model, I suggest reviewing Chapter ten before proceeding. Also, this example is designed more for efficiency and speed of execution over exactness. For example, we will assume linearity (that is, linear behavior between the inputs and outputs) in the interest of keeping the model simple, although linearity does not always hold true (Figure 12.1).

A review of the helicopter and the five variables

In this example, five process variables will be targeted for investigation. As previously mentioned, the experimental model chosen will be the 5-factor, 2-level, 16-run half-factorial. Why not a full-factorial consisting of 32 runs or some other designs? For most industrial optimization experiments, three to five factors will be the norm for investigation. This model can also efficiently analyze six to as many as nine factors. But that will be reserved for advanced reading on that topic. So why a half-factorial? That will be covered next.

A brief background on partial factorial designs and confounding

Partial factorial designs allow for a more efficient design in terms of significant findings in relationship to the number of runs. Table 12.1 represents the typical designs and the resolution of each. There are others, such as Plackett–Burman designs, but they will be saved for further studies in this area.

For an excellent review of the available models, I refer the readers to *Statistics for Experimenters* by Box, Hunter, and Hunter. The main point of Table 12.1 is the understanding of the resolution. Resolution is primarily the risk of a wrong analytical judgment in an experimental outcome without knowledge of steps to mitigate that risk. As will be explained further, the generally accepted rule of thumb for resolutions in order of descending risk levels is as follows:

- Resolution III. Risky as main effects are confounded or confused with two-way interactions. Although commonly left with no choice but to run these designs, they are usually used only for identification of main effects and may require numerous follow-up runs to confirm.
- Resolution IV. Designs are risky at confounding two-way interactions with other two-way interactions, and main effects with three-way interactions. This can be risky, but usually the correct interaction can be determined by identification of a main effect that is associated with the interaction. It is rare that an interaction does not have at least one of the factors as a main effect. This has been found repeatedly by the author and by other experts in the field.

Table 12.1 Available factorial designs with resolutions

No. of runs	Factors						
	2	3	4	5	6	7	8
4	Full	III					
8		Full	IV	III	III	III	
16			Full	V	IV	IV	IV
32				Full	VI	IV	IV

- Resolution V. Main effects are confounded with four-way interactions, and two-way interactions are confounded with three-way interactions. Confounding will be explained further later, but the analysis will return with identical values; thus, with straight analytics, differentiating between the respective outcomes is difficult. The example in the next section is a typical example of a Resolution V.
- Resolution VI and full factorials. These are typically very low risk of confounding significant factors or interactions. For a full-factorial, the risk is essentially zero of confounding unless an extra variable outside the factors to be analyzed is significant (such as ambient temperature). For Resolution VI, main effects are confounded with five-way interactions (rare), three-way interactions with other three-way interactions, and two-way interactions confounded with four-way interactions. All of those occurrences are rare. But Resolution VI and full-factorial come with a price—a relatively high number of runs as shown in Table 12.1.

The 5-factor, 2-level, 16-run half-factorial, Resolution V experiment

For this experiment, there are three primary reasons not to go with the 32-run full-factorial but to instead opt for the half-factorial. In order of importance they are as follows:

- The confounding effects in a Resolution V are low risk. As pointed out earlier, for a Resolution V factorial design, the three-factor interactions are confounded (or sometimes called confused) with two-factor interactions. Is this a legitimate risk for industrial operations? Probably low, as most industrial scientists would argue that three-way interactions are not common. In addition, in a three-way interaction, normally, at least one of the three factors involved in the interaction will be a main factor. If none are, look to the two-way interaction. Bottom line: The additional 16 runs to make it a full-factorial have marginal if any additional benefit.

- We would eliminate many runs that only serve to confirm that one or more factors are benign. After an elimination of one of the five factors, the remaining four represent a four-factor full-factorial. If two factors are benign, there is a three-factor full-factorial replicated (repeated) twice.
- **The overall return on investment is higher than a full-factorial. Performing a 16-run rather than a 32-run is a significant reduction in cost and downtime owing to experimentation. Still, the half-factorial design will frequently result in a significant finding, in either main effect determination or a two- or three-way interaction. Understanding a previously unknown effect on the process is commonly a huge return on the experimental investment.**

The base-level design for the helicopter design

Take five factors from the helicopter design for analysis: wing length, body width, body length, weight, and position on rack (see Figure 12.1). Hold constant the other factors: wing width, intermediate body width and length, wing shape, material type, wing taper, and any other extraneous factors. This will be a two-level factorial design as nonlinearity is not an issue in this opening design. The true skill of the DOE practitioner is not in the analysis but in the next step, developing the table of factors and high–low levels for each factor. This is done as a discussion and typically will

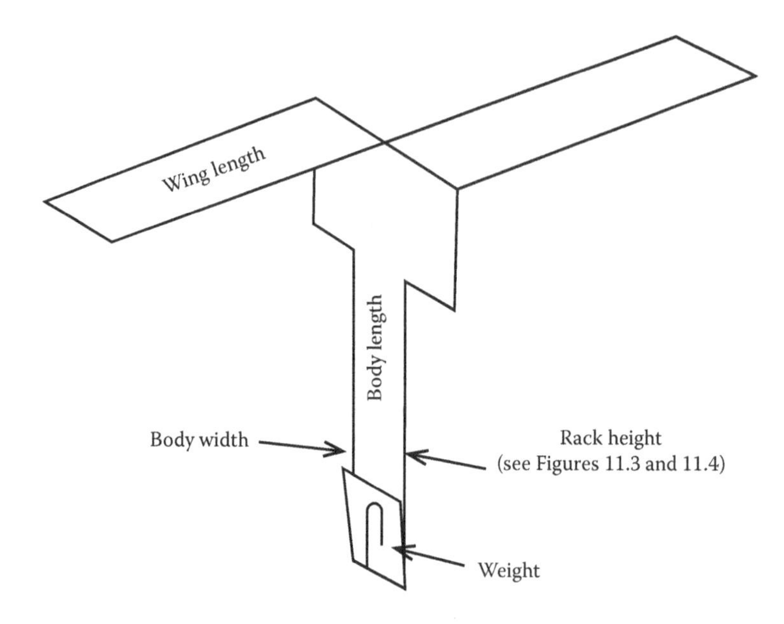

Figure 12.1 Base-level helicopter design.

take between 1 and 4h to determine with line operators and subject matter experts. A successful discussion identifies "the likely suspects" and the magnitude of the low and high levels to ensure that those factors or combinations of factors will have a measurable effect on the experimental outcome without catastrophic effect. (In this experiment, it is known that if the wing length is taken down to zero, it will not fly. This will result in little of any new or useful information. This is what is referred to as a catastrophic effect.) After extensive discussions with the line operators, the decision is to test the following factors at the recommended levels (Table 12.2).

The base-level design will follow the pattern shown in Table 12.3 in coded format.

Table 12.2 The design factors and proposed experimental levels

Factor	Low level (–1)	High level (+1)
Wing length (mm)	70	120
Body width (mm)	35	50
Body length (mm)	70	120
Weight	1 clip	3 clips
Rack position (m)	2.1	2.3

Table 12.3 5-factor, 16-run, half-factorial design coded

Design run	*A*—wing length	*B*—body width	*C*—body length	*D*—weight	*E*—rack position
1	–1	–1	–1	–1	1
2	1	–1	–1	–1	–1
3	–1	1	–1	–1	–1
4	1	1	–1	–1	1
5	–1	–1	1	–1	–1
6	1	–1	1	–1	1
7	–1	1	1	–1	1
8	1	1	1	–1	–1
9	–1	–1	–1	1	–1
10	1	–1	–1	1	1
11	–1	1	–1	1	1
12	1	1	–1	1	–1
13	–1	–1	1	1	1
14	1	–1	1	1	–1
15	–1	1	1	1	–1
16	1	1	1	1	1

The first four variables are a standard format very similar to the DOE in Chapter eight. The last column is composed of multiplying the first four variables ($A \times B \times C \times D$) together in coded format. This results in a design with cube plot of the design shown in Figure 12.2.

The design has several interesting and critical points as follows:

- The design is regarded as orthogonal. Take any factor combinations ($A \times B$, $A \times C$, $A \times D$, $A \times E$, $A \times B \times C$, $A \times B \times C$, $A \times B \times D$, $A \times B \times E$, etc.) and multiply each column; the sum is zero. That is critical for independence of variables or interactions. Also, all main effects columns add to zero.
- Notice from the earlier point that if one factor is eliminated, what remains is a full-factorial on the remaining four factors. This can best be seen visually by looking at one of the factors independently. Looking at Figure 12.2 factor *E* from the top two cube plots (the +'s) to the bottom two cube plots (the –'s), all corners of every cube are included. Thus, eliminating that factor would result in a full-factorial of the remaining four factors. That is a huge bonus when running experiments as at least one factor is normally found benign.
- Factor *E* in the fifth column has as a special characteristic. This factor is more subject to confounding than the other factors. If engineering knowledge is available, this should be the least likely factor to interact with the other variables. This may generate the next question:

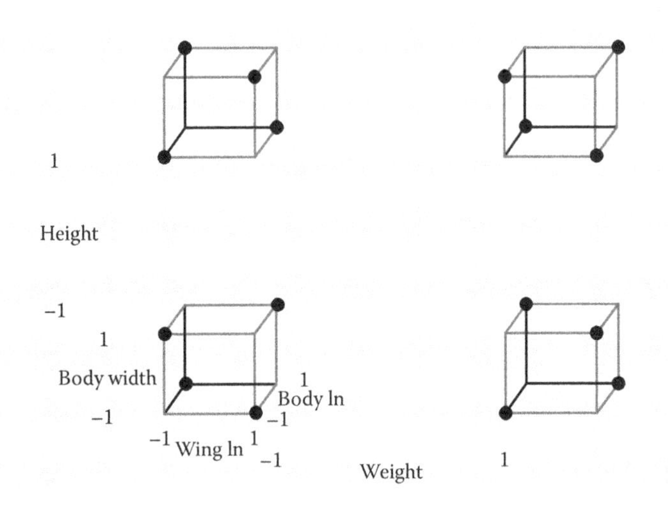

Figure 12.2 Cube plot of 5-factor, 2-level, half-factorial.

If that is known, why would it be necessary to run the experiment? Valid point, but notice the language—least likely to interact. If it is a primary factor and acts on the process independently, this experiment will find it. But numerous times, a certain variable will not likely physically interact with the other variables in the experiment. (Example: most commercial wood sanding operations: from the author's experience, running multiple DOEs in multiple similar operations, the paper grit is an independent variable.) If that is the case, placement of this variable in the last column will minimize confounding.

- Sixteen runs in many facilities are hard to organize and run. If there is any doubt about the integrity or about the mixing of runs, either rerun those runs or drop them all together. Normalization of one run is better than a major error in judgment based on sloppy record keeping.

The results and analysis

As with the prior experiment in Chapter ten, randomization is probably even more critical in this experiment. The opportunity for confounding one or more factors with an environmental factor such as the ambient temperature or with a raw material is greater than that with the eight-run experiment. Here is an easy method of randomization for experimental runs.

From a deck of cards, take cards ace (1) through 10 plus the jack, queen, and king of the same suit. For this example, utilize hearts. Combine that with the 4, 5, and 6 of another suit. For this example, we will use diamonds. Let the ace through king of hearts represent runs 1 through 13, and the 4, 5, and 6 of diamonds represent runs 14-15-16. Take the 16 cards and shuffle a minimum of four times. Place the cards in order of the shuffle from left to right on the table. Run the experiment via the card order on the table. Example: After the shuffle: 4H, 9H, 5D, AH, 8H, KH, QH, 4D, 3H, JH, TH, 2H, 6D, 7H, 5H, 6H. Design run 4 first, followed by design run 9, followed by design run 12, etc.

The final design and typical results of the experimental runs are presented in Table 12.4.

A nonanalytic graphical view of the experiment

> It is not unusual for a well-designed experiment to analyze itself
>
> Dr. George Box

Table 12.4 Results of the experiment—replication seven per run

Design run	Actual run	*A*—wing length	*B*—body width	*C*—body length	*D*—weight	*E*—rack position	Flight time (s)
1	4	−1	−1	−1	−1	1	1.4
2	9	1	−1	−1	−1	−1	2.1
3	15	−1	1	−1	−1	−1	1.3
4	1	1	1	−1	−1	1	2.2
5	8	−1	−1	1	−1	−1	1.3
6	13	1	−1	1	−1	1	2.0
7	12	−1	1	1	−1	1	1.1
8	14	1	1	1	−1	−1	2.3
9	3	−1	−1	−1	1	−1	1.1
10	11	1	−1	−1	1	1	1.5
11	10	−1	1	−1	1	1	1.3
12	2	1	1	−1	1	−1	1.4
13	16	−1	−1	1	1	1	1.4
14	7	1	−1	1	1	−1	1.3
15	5	−1	1	1	1	−1	1.3
16	6	1	1	1	1	1	1.4

As Chapter eleven highlighted, there are three primary keys to effectively implementing DOE. They are as follows:

1. Reduction in variation of all variables throughout the test. This is critical and may consist of a few of the following extraneous factors held constant: raw material, operator, testing equipment, gauges, environmental conditions such as humidity or ambient temperature, line processes such as hydraulic pressure, etc.
2. Maximizing data in the output parameter (flight time in this example), as a result of a change in the input parameter. In other words, make sure the input parameters are moved sufficiently to see the change in the output. For this experiment, the wing length is moved from 70 mm to 120 mm, which is very likely to show up in the outcome if it is a factor. Movement from 90 mm to 95 mm will likely result in little, if any, movement and the eventual incorrect conclusion that wing length was benign (a major beta error).
3. Organization. A change in the run order, mislabel of the data, and unplanned reduction in the number of tests for each run are just a few of the problems with poorly executed DOEs. A well-designed DOE with a poorly executed test can ruin the results and lead to beta error but probably more important, a disastrous and embarrassing alpha error.

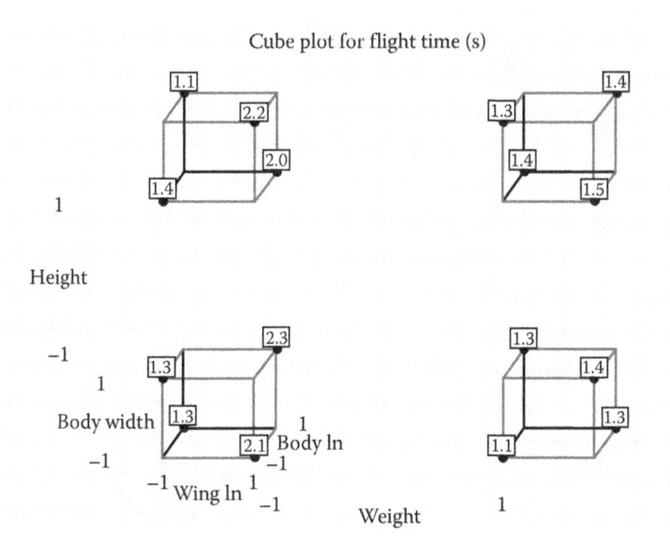

Figure 12.3 Cube plot of helicopter experiment.

The typical results of the aforementioned DOE are as shown in Figure 12.3.

Before proceeding, suggest pausing and viewing the graph in Figure 12.3 in detail. From the earlier, there are several basic findings:

- There is a significant interaction between the wing length (*A*) and the weight (*D*). This section will be shown in more detail later.
- The height, body width, and body length are benign. That may not seem intuitively correct, but the data do not support intuition. Any change in the process centered on those three variables will likely result in no change in the process. In practice, any use of capital to change those processes with the expectation that it will change the results will likely result in frustration and disappointment with the entire program. This inadvertently leads to changes in the entire direction of the continuous improvement system or program initiated.
- The interaction term may have been missed without the use of a designed experiment, and at a minimum, it would have been suboptimized.

Eliminating the three benign variables results in the interaction as shown in Figures 12.4 and 12.5 in graphical form.

(For further discussion regarding the construction of interaction plots, see Box et al. 2005, pp. 173–193.)

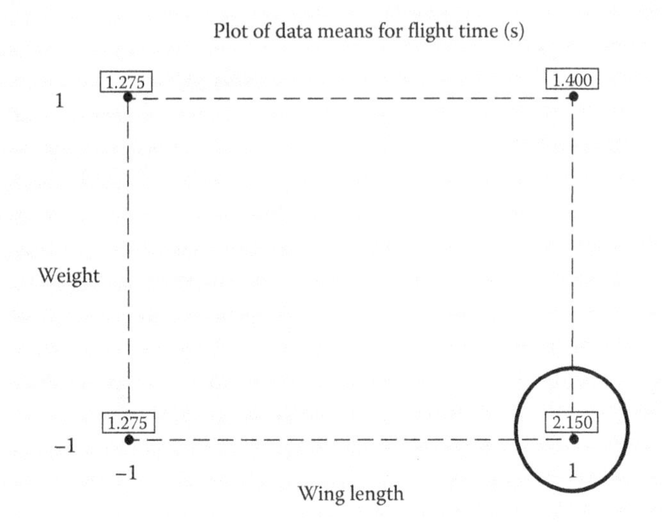

Figure 12.4 Plot of flight times by only two critical factors found—weight and wing length.

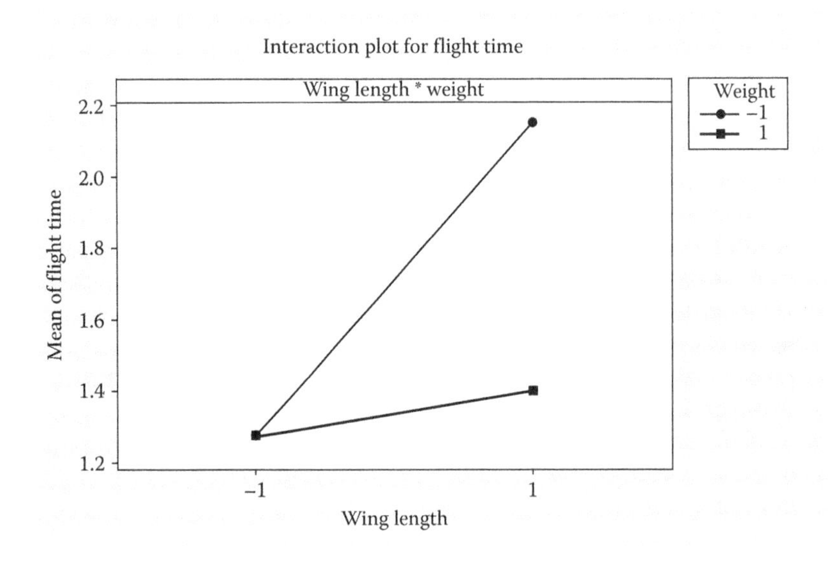

Figure 12.5 Interaction plot of wing length and weight.

Typical next steps

The next steps will be completed in chapter thirteen. Is this the stopping point in any experiment? Hopefully not, as these results provide only a direction for potential major improvement and not an ending point or optimization. Nor does this represent the opportunity for potential reduction in cost of the benign variables. The initial DOE answers the question, "What factors are important to the outcome of this process?" The follow-up can then address what to do to optimize the outcomes and minimize costs.

Discussion areas and potential next steps for students

- What is a similar experiment that could be run for the juggling example in Chapter ten? What factors were potentially missed in that DOE? Take this next step and implement the DOE with two new variables—ball diameter and starting hand. The results may be surprising.
- What do you recommend as next steps for the benign variables, body width, and body length? What if these are high-cost variables? What might be the restriction on lowering the cost?
- Discuss the ramification of not finding the measured height difference as significant. How could that be possible if there was in fact a difference in measured height? What is the proper answer to management that states that there must be difference, and I can see the height difference!
- Brainstorm for other 5-factor, 2-level, 16-run DOEs on existing processes not necessarily in manufacturing. Example 1: Automobile mileage as the outcome and experimental factors: tire pressure, driver, driving method—city or expressway, fuel octane, dirty or clean air filter. Example 2: Bicycle time from point A to point B: factors might be seat height, tire pressure, handlebar height, breakfast that day (yes/no), water intake during ride (yes/no).
- For manufacturing. Arrange a plant trip and analyze each manufacturing process. What is the scrap rate? What is the speed to design? What is the downtime? Describe a DOE to reduce any of those factors.
- For nonmanufacturing. Describe a designed experiment for a nonmanufacturing process such as customer service response. What are the similarities to the manufacturing DOE? Discuss how you would run the DOE and issues associated with it.

- For historical data. What methods that follow the DOE process can be utilized for historical data analysis? Hint: The 2008 Presidential election may have been decided by the use of this method.
- How does this process work for a start-up operation? What differences might there be in the method variables identified for analysis in the DOE? What differences are there potentially in the number of variables?

Bibliography

Anderson, Virgil L.; McLean, Robert A., *Design of Experiments: A Realistic Approach*. 1974, Marcel Dekker, Inc., New York.

Box, George E.P.; Stuart Hunter, J.; Hunter, William G., *Statistics for Experimenters: Design, Innovation, and Discovery*, 2nd Edition. 2005, John Wiley & Sons, Hoboken, NJ.

Moen, Ronald D.; Nolan, Thomas W.; Provost, Lloyd P., *Improving Quality through Planned Experimentation*. 1991, McGraw-Hill, New York.

Montgomery, Douglas C., *Design and Analysis of Experiments*, 5th Edition. 2001, John Wiley & Sons, New York.

chapter thirteen

Process optimization after the design of experiments

Objectives

- Understand typical process optimization steps after running the design of experiments (DOEs).
- Walk through a hands-on example of the optimization process with the experimental helicopter.
- Review some of the common what-ifs when applied in an actual application.

Overview from the 5-factor, half-factorial, designed experiment

The last chapter left off with the finding of a major interaction between the wing length and weight. Interactions between two or more factors are common in most industrial processes. They are also the source of common frustration for operators. Statements from frontline operators such as "this process has a mind of its own—1 day it works fine, the next day it will not" are not unusual in processes not optimized. Why? Because it is very difficult for any operator to pick up on a process interaction without using a systematic experimental approach such as DOE. In Chapter twelve, the DOE located the underlying cause of the frustration—this next step will optimize it further, potentially reducing the underlying causes or, in some cases, eliminating them completely. This optimization process is unique to the operation and should not be utilized or interpreted as a standard to follow as there are multiple different excellent methods. Three typical follow-ups to any DOE are covered.

The optimization process with the experimental helicopter

There are typically three phases to the follow-up process. Follow-up testing to determine optimal settings of the critical factors, followed by confirmation of the optimized process, and then actual implementation or what is commonly called "go live" or implementation in the actual operation.

Of the three, the most difficult but rewarding is the "go live" phase. This will be discussed in detail later.

The follow-up testing

What was found after the last DOE was a two-way interaction. Drawing only that interaction results in what is shown in Figure 13.1.

As shown in Chapter twelve, there is a large interaction between wing length and the weight (Figure 13.1 at wing length +1 and weight at –1). At this point, if the results have been reported to management, there is likely a drive to implement changes as soon as possible, especially if the process was chosen for the DOE because of poor quality and/or high cost. There are also likely limited resources remaining to experiment with the process. The typical rule of thumb is to allow 25% of the experimental budget on the first DOE. But scarce funds for experimentation and equipment downtime are mostly difficult to procure. How best to optimize after the likely costly opening DOE and the drive to implement without optimization?

The problem with the abovementioned outcome is the factors tested may not be linear. Eight more runs will likely locate close to an optimum point when limited resources are available for an extensive follow-up optimization. Although there are multiple options, a typical pattern follows as shown in Figure 13.2.

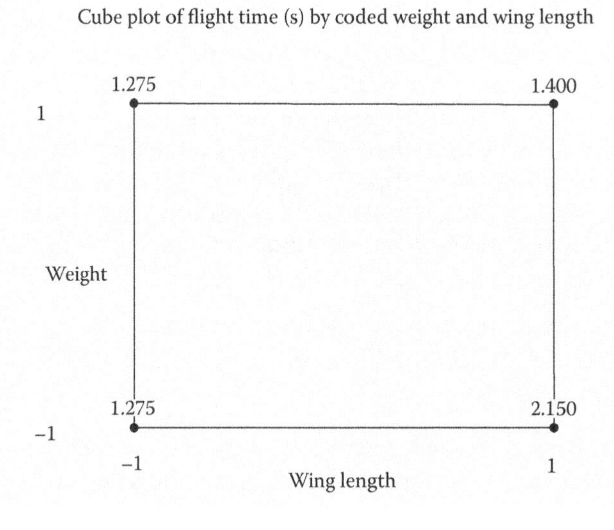

Figure 13.1 Flight time and weight interaction in coded format.

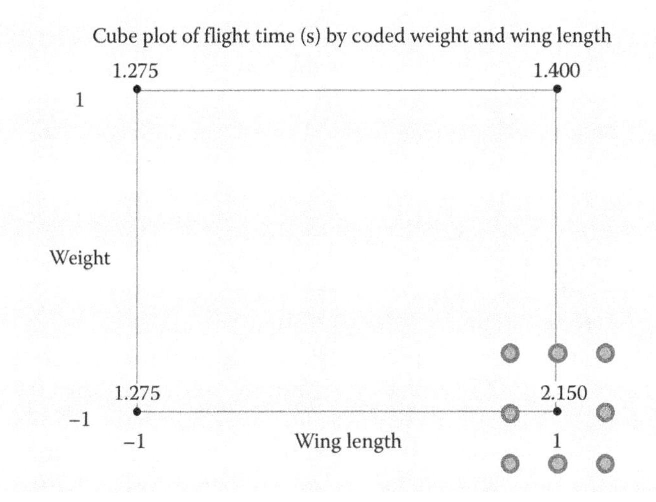

Figure 13.2 Experimental runs between wing length and weight.

This is a typical follow-up to the first experiment and only requires eight more runs with all other factors (wing width, body width, body length, paper type, etc.) held constant. Why this pattern? This will determine an optimization location with a minimum number of runs. Notice this pattern is a 3-level, 2-factor, full-factorial run in 9 runs without the center point, which was run in the base DOE in Chapter twelve. This can be computed as follows:

$$\text{No. of runs} = \text{No. of levels}^{\text{No. of factors}} = 3^3 = 9$$

The width of the pattern depends on the economics of the remaining runs. In most real manufacturing processes, this may consume the majority of the project budget. Finding micro options outside of the abovementioned points found in the prior experiments may not even be an option.

Before running this follow-up testing, run the center point first to reconfirm the results obtained during the initial DOE. If the mean and/or standard deviation are significantly different from that when that test point was run during the DOE, stop the process until determining the root cause or an unaccounted factor. In an actual manufacturing process, this step is easily missed and can result in a missed call on the process at the implementation phase.

An easy method to teach this confirmation step is with the experimental helicopter and rack system. Have the continuous improvement candidates, or students, run the center point first. Between the original DOE and the confirmation run of the center point, change the height of the rack used to drop the experimental helicopter. This is an advantage of utilizing the adjustable racks and not the typical method of the fixed point such as the ceiling. When dropping the two and comparing them with a 10- to 15-cm difference in height, there should be a significant difference in flight time. The typical output would be similar to that shown in Figure 13.3.

This should be a struggle for students but not uncommon to real-life occurrences on an actual manufacturing or nonmanufacturing operation. What changed? Why was the process potentially optimized yesterday and not today? Was it the raw material? What batch number was used yesterday? Was it a different operator, and if so were the settings changed? Was there a preventative maintenance performed on the machine overnight? Did the environment change significantly? If a hidden factor was the humidity, did a cold front come through overnight? Was there a fan on yesterday during the test? Has the experimental helicopter design changed because of batch-to-batch differences in design? What about the mold? Was it changed overnight? If this is now the weekend or a nontypical operating day, did the power factor change? If this is a nonmanufacturing process, was there an incoming product change such as a surge in customer demand in a bank or retail outlet or customer service facility? Was there a change in measurement systems? Was a new gauge operator started on the process?

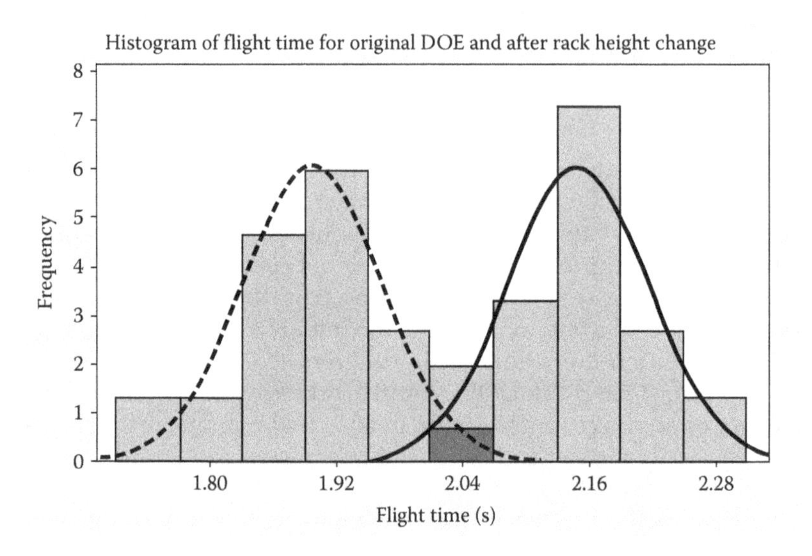

Figure 13.3 Optimization at the completion of the DOE and after the rack change.

After struggling with the investigation, check the height and become aware of the difference. A height difference of 10 cm is unlikely to be seen visually, especially without a comparison from the prior test. At the completion of the investigation, have the students recalibrate the height and run another test to confirm. There should be no significant difference between this run and the optimization point found in the DOE.

The frustration in that process will be a small example of what is required if this happens in a real operation—and most likely it will. Also notice the effect that the concepts outlined in prior chapters have on this investigation: The lower the variation, the more likely a difference will be found. The aforementioned difference may not have been found if all the other factors are not blocked during the follow-up phase and eventual implementation. Without knowledge of the hypothesis experimental method and the application of the base-level fundamental method of it, the difference would most likely not have been found.

After confirming the center point, obtain data for each of the remaining eight runs in the follow-up testing. Enter the data into a graphical format or a statistical software package such as Minitab, JMP, or SPSS.

Interpretation and confirmation of the new optimization point

The optimization point is between tested points and can be seen from the results in the two graphs (Figures 13.4 and 13.5). Although the two graphs can be constructed by hand, for introductory DOE learning, these are best constructed in the statistical software packages identified earlier and typically used in tandem. The two graphs basically reach a similar conclusion: The optimum point can now be found based on capability levels and cost. In the hypothetical model, the optimum operating settings will depend primarily on two factors: The cost of the added wing length, the benefit, and other potential negative effects from the weight removal. In a real manufacturing operation, there will likely be additional factors—for example, in the plant making paper clips or pace makers; what is the risk level of a part manufactured outside the customer specifications?

For this analysis, with only eight new points, the optimum point is located at a coded wing length of 1.1 and a coded weight of –1.1.

A brief explanation of coded units

It is best to utilize coded units in designed experiments. Why? As will be shown in the following paragraphs, interaction terms are difficult to

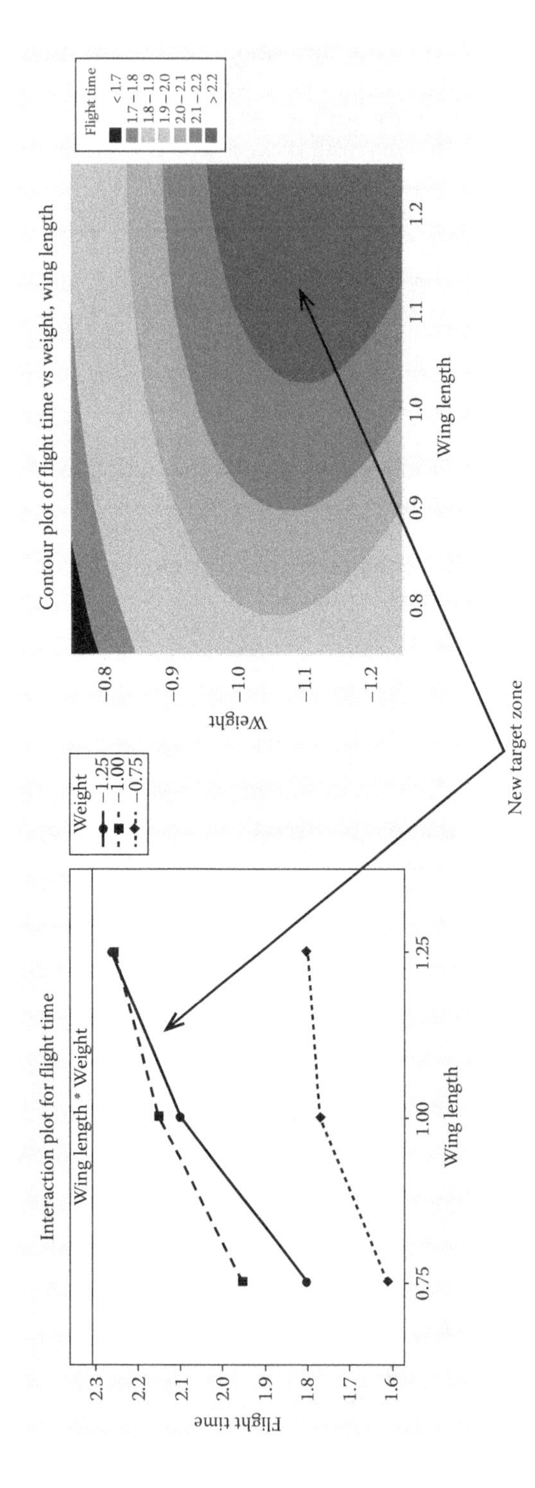

Figure 13.4 and 13.5 Showing optimization location.

utilize in regression models if not in coded units. To understand the derivation of coded units, think in terms of a sliding scale based on the original two levels determined in Chapter twelve (Figure 13.6).

To derive a number (X) on the scale:

If above 95,

$$\frac{X-95}{120-95}$$

If below 95,

$$\frac{X-95}{95-70}$$

So as an example 130 mm in coded units would be,

$$\frac{130-95}{120-95}=1.4$$

For this example, compute the wing length and weight from the coded units as follows (Figure 13.7):

Wing length:

$$\text{Coded unit } (1.1)=\frac{\bar{X}-95}{120-95}$$

$$\bar{X}=125.5 \text{ mm}$$

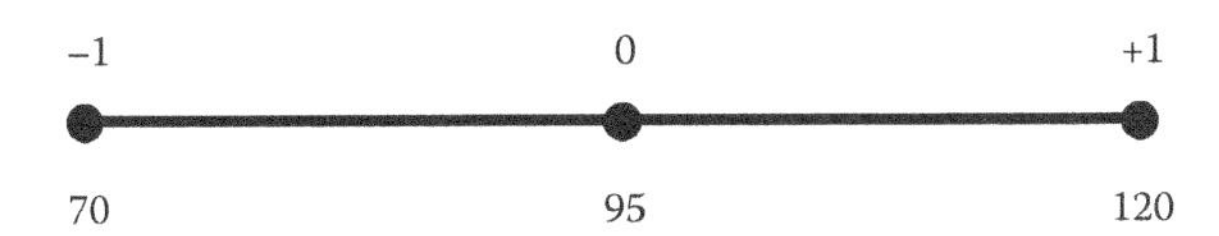

Figure 13.6 Coded units scale for wing length.

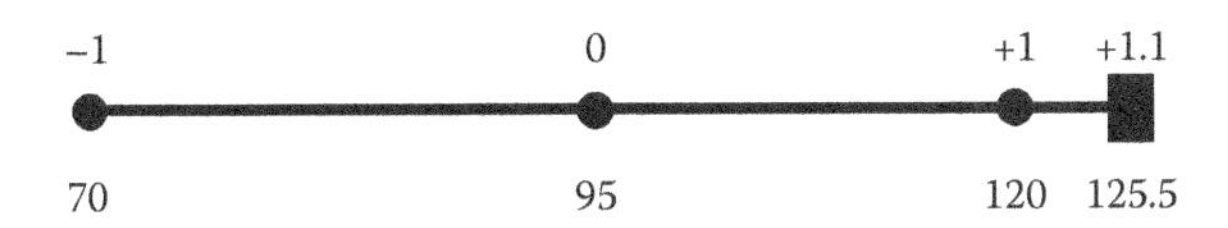

Figure 13.7 Coded units for wing length of 1.1.

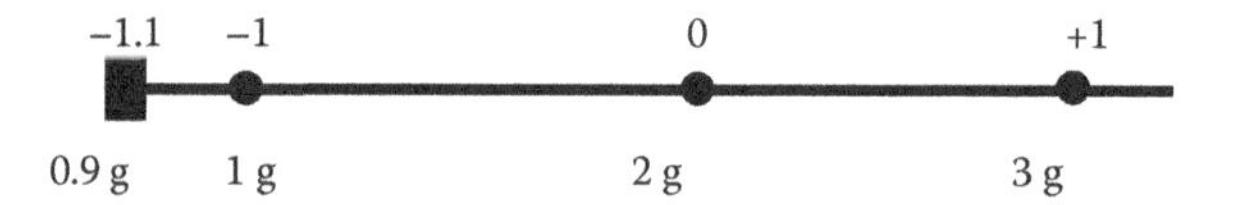

Figure 13.8 Coded units for weight (g) for –1.1.

Weight: (note standard weight of the zero point is 2 g, –1 coded units is 1 g, +1 coded unit is 3 g) (Figure 13.8).

$$\text{Coded unit}\,(-1.1) = \frac{X-2}{2-1}$$

$$\bar{X} = 0.9\ \text{g}$$

Now that we have our optimized helicopter settings, what flight time can be expected? This can also be calculated, using the regression model. Regression models can be developed for this scenario from the DOE by hand, by formula, or from the statistical software packages referred to earlier.

Regression model with interaction term added in:

$$\text{Flight time} = 1.52 + 0.25 \times (\text{Wing length}) - 0.19 \times \text{Weight} - 0.19 \times \text{Wing length} \times \text{Weight}$$

Substituting in the coded units for Flight time (1.1) and weight (1.1)

$$\text{Flight time} = 2.2\ \text{s}$$

Note that we have found the optimum values and expected flight time using experimentation and calculation, without actually needing to run a physical experiment at the optimum values! Using the DOE and coded values, we have gathered several points and then interpolated the best value. When possible, confirm the results by running the process with the selected factor values to see if the expected outcomes hold. How does the optimized setting compare to the previous settings before the DOE? It's not uncommon to find a massive improvement.

Utilizing the existing standard deviation (0.15) and the minimum specification of 1.5, the new capability to the lower limit (CPl) is as follows:

$$\text{Capability to the lower limit (CPl)} = \frac{\bar{X} - \text{Lower spec}}{3 \times \text{Standard deviation}} = \frac{2.2 - 1.5}{3 \times 0.15} = 1.55$$

Putting it into action or "go live"

The previous section demonstrates the sometimes difficult steps of optimizing a process. The process will not likely improve without experimentation and will only get worse by adding inspectors or requesting the operator to adjust when it gets close to an out-of-specification condition. Optimizations will work, but it takes effort and time. Learning how to optimize by this methodology will improve almost any process.

For an actual operation, the next step is to implement in operations. That can be time consuming and frustrating, as issues undetected before will undoubtedly surface. Comments such as "we never had this happen before" are common but expected because the new process is running with different input parameters. Once in operation, expect the process to improve and then optimize with further microexperimentation commonly called evolutionary operations.

"Go live" phase usually is accompanied by a process of mistake-proofing of the critical input parameters. Notice now, the parameters are optimized for the wing length and weight. Since these are critical to the performance of the helicopter meeting the 1.5-s customer requirements, these require the highest level of quality and assurance of meeting customer specifications. Typical mistake-proofing techniques for this example might include a device at the end of the line to prevent a condition of over or under size. For the weight, it might include the operation of an automatic weight scale and shutdown system if out of specification. These are systems to find mistakes after the fact, but the best assurance is to prevent an oversized wing or overweight condition from ever being produced. This might include assurance techniques during the molding process or other areas at the material supply process.

Just as important to the control phase is what is not required for the other factors found benign, such as body width. Expensive devices are not likely required for this parameter as the DOE and optimization process found are not critical to the characteristic of flight time. A check of the process for manufacturing the body width may be necessary but only for a critical cost range. If so, a simple control chart checking every 10th part may be sufficient, but a major investment is likely not needed. Notice the parameters requiring control are determined from experimentation and not from "gut feel" or "past experience." The data rules the process and determines what needs controls—not the other way around.

Follow-up exercises for students

- This experiment was designed to find factors that potentially have an effect on the mean of the process. Design another experiment to optimize the standard deviation. What critical factors control the standard deviation?

- For a final flight test, drop another series around the factor or factors for the raw material. Add three different materials for experimentation. What does the output determine? Why is it critical? What does that tell you about the requirements for production and supply management?
- After optimization, utilizing the rack and experimental flight, take the model and drop until material failure. If the abovementioned material tests results using a thinner or lighter material, what effect does that have on the cycle to failure?

Bibliography

Box, George E.P., Response Surface Methods: Some History, Chapter 2. In Blading, David et al. (eds), Editors Emeriti: Barnett, Vic, Hunter, J. Stuart, Kendall, David *Improving Almost Anything: Ideas and Essays*, Revised Edition, 169–171. 2006, John Wiley & Sons, Hoboken, NJ.

Box, George E.P.; Draper, Norman R., *Empirical Model-Building and Response Surfaces*. 1987, John Wiley & Sons, New York.

Moen, Ronald D.; Nolan, Thomas W.; Provost, Lloyd P., *Improving Quality through Planned Experimentation*. 1991, McGraw-Hill, New York.

part four

Data, statistics, and continuous improvement for everyone

chapter fourteen

Making data, statistics, and continuous improvement fun and interesting

A typical week-long course

Objectives

- Develop a first-step hypothetical syllabus for a classroom Six Sigma course.

An overview

This book has described an environment of education that makes learning fun and interesting. Taking many courses that historically have been traditionally taught, this approach adds the interesting factor to the learning methodology. It is hoped that students learning by this approach for the first time will find data, statistics, and continuous improvement more interesting, leading to more students engaged in these fields of study.

Typical introductory Six Sigma or continuous improvement education model with this approach

Day 1: AM

Introduction to mean, standard deviation, confidence interval, and basic statistics utilizing the normal distribution by juggling as outlined in Chapter six. By the end of this section, most students will start to understand many areas including one critical concept—this is not a normal class. They should finish this day mentally excited and physically exhausted.

Day 1: PM

Introduction to hypothesis testing by experimental helicopter as outlined in Chapter eleven. This will be the introduction to this concept, critical for understanding beta error, alpha error, and power. This is an excellent time to introduce Deming's concept of overadjustment or the funnel experiment. New to most students, managers, and executives, hypothesis testing can have a very powerful long-term impact on business, government, or any micro/macro process as highlighted in Dr. Deming's books on data and continuous improvement.

Day 2: AM and PM

Probability, including the multiplication principle by card manipulation, introduction to Bayesian statistics by coin magic, binomial probability function, and (not covered in this book) the data for rare events utilizing the Poisson distribution. Card manipulation and coin magic techniques are described in Chapters two through four.

Day 3: AM

Statistical process control (SPC) by juggling and experimental helicopter. As reviewed in Chapter seven, students can develop an SPC chart for the tosses to drop, determining whether their own process is in a state of control from morning to evening. As a possible addition, develop an SPC chart for the helicopter drops at 2.5 m. As an added special effect, turn on a fan half way through the drops to introduce an environmental factor. This is an excellent way of introducing the power of SPC for identification of special causes of variation.

Day 3: PM

Gauge R&R by the experimental helicopter, not covered in this book. With the addition of the adjustable rack, construct a test ball to drop from the rack height to floor level, removing the environmental effects of the experimental helicopter. With students split into teams, a traditional Gauge R&R study will review shortcomings in the timing system (the stop watch) by means of the process-to-tolerance ratio and analysis of differences between operators. This will challenge students to develop different techniques to remove the noise in the gauge. This is typically very similar to an actual manufacturing or transitional process.

Day 3: PM

The process capability index and the meaning behind it by experimental helicopter and/or juggling. As covered in Chapter six, the process capability index will determine the required continuous improvement path—centering, improvement by control, variation reduction, or a combination of all three. This can be done by two methods. The first is utilizing the experimental helicopter with the lower specification at 1.5 s for a 2.5-m drop. This will drive the improvement process as the capability, if in control, will be significantly <1. For the juggling process, the lower specification will be determined by the student. The lower specification will be determined by their own desire. This may be a minimum of 5 to, in some cases, 10 tosses to failure as the desired minimum specification. Again, this is up to the client—in this case, the students themselves.

Day 4: AM and PM

Design of experiments (DOEs) by experimental helicopter, juggling, and card magic. Start with a very basic juggling DOE as described in Chapter eight. What factors are critical? Follow up with new analysis of control and capability index. Continue with the experimental helicopter, 3-factor, 2-level, 8-run design. Easy to do, and a very obvious two-way interaction. Finish with the 5-factor, 2-level half-factorial as described in Chapter twelve. Along the way, the card trick of the 5-factor, 2-level, 32-run full-factorial DOE from Chapter eleven is an excellent way to explain the concept of orthogonality. By the end of this day, all students should have a midlevel knowledge of DOE and be able to apply it to a manufacturing or transactional process.

Day 5: AM

Regression and optimization. Chapter thirteen describes the optimization process after the 5-factor designed experiment with the experimental helicopter. This is an excellent way to find an optimization point and demonstrate response surface methodology. At the completion of this exercise, the regression model as demonstrated in Chapter nine with the juggling process is a way to find an optimum CPl. At the completion, each helicopter team can build their own custom-designed experimental helicopter with their settings for weight, wing length, body width, and body length showing evidence of SPC and a new optimum CPl.

Day 5: PM

Testing for long-term control and the bathtub curve as outlined in Chapter eight. This final exercise is to test out the bathtub curve for the experimental helicopter–optimized design. How many cycles does the curve transition from the constant failure zone to the wear-out zone? Attempt this with multiple helicopters. Is the point predictable? If so, and the maintenance practice is for a 1% downtime rate, at what point should the helicopter be taken out of service?

A course developed on statistics and probability that is hands-on is going to be much more adaptable to the world of the future as the information age expands. This concept will be explored more in the final chapter.

Extra hands-on exercises

- Take the experimental helicopter and change the material utilizing three different materials: thin gauge paper, heavy cardboard paper, and paper thin plastic. What changes with the characteristics? Is there a significant improvement or no changes? What happens to the overall weight and the cost of operations?
- Take the three raw materials and determine the transition point from constant breakdown to wear-out. Does this point change?
- Have all students take a process from outside the classroom and optimize it through the aforementioned process. Trouble coming up with a process? Here are a few:
  - Baking bread
  - Time to ride a bike from point A to B—this is a classic 7-factor DOE
  - Their automobile gas mileage
  - Flying a kite—what are they trying to optimize?
  - Their household BTU consumption
  - Their household water softener usage
  - Their blood sugar level
  - Their cholesterol level.

chapter fifteen

Final thoughts

Properly utilized, this approach will make the learning process for data and statistics interesting and even fun. This could potentially create a complete generation of new data, statistics, and probability specialists.

In 2006, Salman Khan, creator of the Khan academy, gave a Ted Talk on the future of Mastery Learning. In that talk, Sal Khan described two triangles representing the population by skill level now and in the future. The present outcome of the system of learning can be characterized by the first triangle or pyramid—a large pool of human labor, largely uneducated in the sciences of data, statistics, and advanced continuous improvement techniques. The middle section is characterized by bureaucrats and information specialists. The third section is researchers, creators, and continuous improvement fanatics leading to entrepreneurship and ownership of capital. What a tragedy of lost potential!

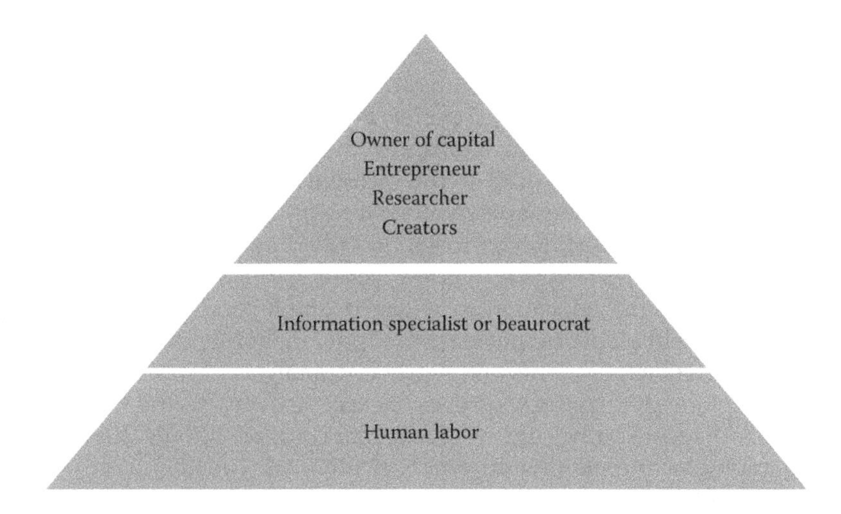

But what if the model was reversed, which is probably happening throughout the world? Human capital and bureaucrats are being replaced by computers and robotics. Creators, researchers, and entrepreneurs are now the top tier. Breakthroughs in new technology or medical research previously never allowed to happen are now commonplace. A person who was previously a laborer in a machine shop sweeping floors now

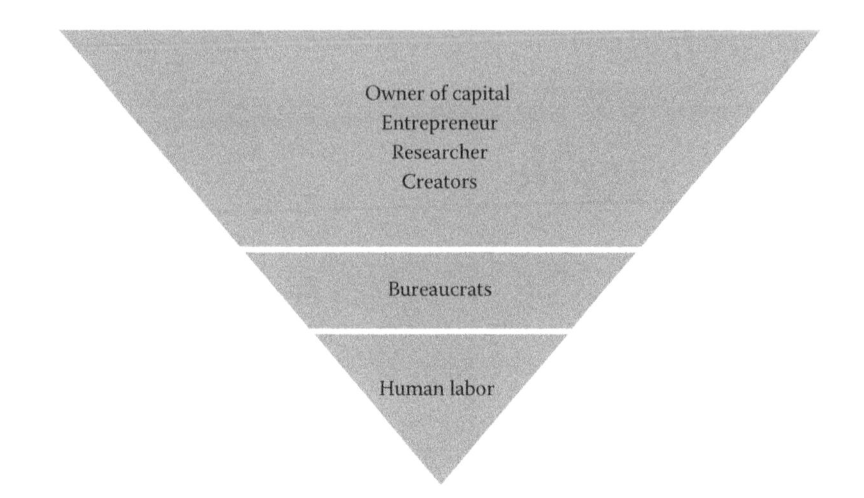

becomes a continuous improvement specialist, proving that a different speed and cutting angle in tangent can extend the life of machine tooling by a factor of 3 to 1. A grinding wheel operator can find a way to eliminate the incoming defect instead of grinding the burr off the part. That end-of-line inspector now finds a way to completely eliminate the incoming part defect.

Utopia? Maybe not. How many students, born into less than ideal education systems, failed to become excited by the concept of statistics, data, probability, and continuous improvement? How many became disenchanted because of the way it was presented or taught? How many entrepreneurs, researchers, and creators are waiting to be found?

The author has used the approach described in this book throughout the world. This process has converted workers and frontline employees with limited formal education to engaged employees who understand basic concepts in statistics and data analysis, including many who now utilize design of experiments and regression analysis.

To explore this concept further, please visit my website at www.williamhooperconsulting.com, research other concepts through the chapter bibliography, or pursue this approach at other forward-thinking educational institutions such as the Kahn Academy at www.kahnacedemy.org.

Index

S

T

U

W